# Solar Cells - New Aspects and Solutions

# Solar Cells - New Aspects and Solutions

*Editor*

Lucas Formann

scitus academics

**Solar Cells - New Aspects and Solutions**
Edited by **Lucas Formann**

ISBN: 978-1-68117-614-7
Library of Congress Control Number: 2016946472


616, Corporate Way, Suite 2, 4766,
Valley Cottage, NY 10989

www.scitusacademics.com



**Notice**



Printed in United States of America on Acid Free Paper

# Preface

A solar cell (or a "photovoltaic" cell) is a device that converts photons from the sun (sunlight) to electricity. In general, a solar cell consisting of non-solar and solar light sources (such as incandescent bulb photons) is called a photovoltaic cell. In essence, the unit must perform only two functions: the photographic generation of charge carriers (electrons and holes) in a light absorbing material and the splitting of the charge carriers into a conductive contact that can transmit power. This transformation is recognized as the photovoltaic effect, so that photovoltaic is the unit-related research area. Solar cells have legitimate uses. Units have historically been used in cases where power is not available, such as energy systems in rural areas, Earth - orbiting satellites, consumer systems, for instance. Pumping applications for portable calculators or possibly wrist watches, telephones and radio. Solar cells are usually considered among the emerging technologies for sustainable energy.

This book contains up-to-date information particularly related to the photovoltaic generations, new innovative ideas and technical solutions of solar cells.

# Table of Contents

# Chapter 1

## Porous (001)-faceted anatase $TiO_2$ nanorice thin film for efficient dye-sensitized solar cell

**Athar Ali Shah, Akrajas Ali Umar[a] and Muhamad Mat Salleh**

Institute of Microengineering and Nanoelectronics, Universiti Kebangsaan Malaysia, 43600 UKM Bangi, Selangor, Malaysia

## ABSTRACT

Anatase $TiO_2$ structures with nanorice-like morphology and high exposure of (001) facet has been successfully synthesized on an ITO surface using ammonium Hexafluoro Titanate and Hexamethylenetetramine as precursor and capping agent, respectively, under a microwave-assisted liquid-phase deposition method. These anatase $TiO_2$ nanoparticles were prepared within five minutes of reaction time by utilizing an inverter microwave system at a normal atmospheric pressure. The morphology and the size (approximately from 6 to 70 nm) of these nanostructures can be controlled. Homogenous, porous, 5.64 ± 0.002 $\mu$m thick layer of spongy-nanorice with facets (101) and (001) was grown on ITO substrate and used as a photo-anode in a dye-sensitized solar cell (DSSC). This solar cell device has emerged out with 4.05 ± 0.10% power conversion efficiency (PCE) and 72% of incident photon-to-current efficiency (IPCE) under AM1.5 G illumination.

# 1 INTRODUCTION

$TiO_2$ nanostructure with a larger surface area is ideal for solar cells [1, 2, 3], photolysis [4], sensors [5] and photocatalytic applications [4, 6, 7], as it improves the charge-transfer reaction, enhances the redox potential of photogenerated electrons and holes, and reduces the electron-hole recombination. For a solar cell application, $TiO_2$ with large surface area provides many active centers for reagent adsorption and reaction, improves dye molecules loading and facilitates facile electrolyte diffusion, leads to a facile electron transport in the device [8, 9, 10, 11, 12].

In a dye-sensitized solar cell (DSSC), anatase is the $TiO_2$ polymorph that shows an intriguing performance [13, 14, 15, 16, 17]. Since many surface reaction favours to occur at the high-energy site, such as defect, twinning or kinks [18, 19], to synthesize anatase $TiO_2$ nanostructures having such structural properties promises enhanced performance in applications. Moreover, anatase $TiO_2$ with high-energy plane, such as (001), and anisotropic-shape (such as nanorice) [20], containing high-surface area and high-defect further promotes active surface reaction and facile electron transfer in the device [5, 21]. Thus, high-performance solar cell or photocatalysis can be obtained from the structure.

In this paper, we present a straightforward method to prepare anatase $TiO_2$ nanorice with a large-area of high-energy plane of (001) containing high-surface defect via a microwave assisted liquid-phase deposition method. In typical procedure, $TiO_2$nanorice (size in the range of 6 to 70 nm) with high-density (thickness of approximately 5.64 ± 0.002 $\mu$m) can be successfully grown on an ITO substrate surface via this method using a growth solution containing $TiO_2$ precursor and hexamethylenetetramine (HMT). The performance of the $TiO_2$ nanorice in DSSC has been examined. Power conversion efficiency and incident photon to current efficiency as high as 4.05 ± 0.10% and 72%, respectively, can be achieved so far. The performance of the device could be further enhanced via $TiO_2$ nanostructure properties as well as device properties improvements. The porous $TiO_2$ anatase should find a potential used in solar cell and photocatalysis applications.

# 2 EXPERIMENTAL

## 2.1 Synthesis and characterization of $TiO_2$ nanorice on an ITO substrate

The $TiO_2$ nanorices were synthesized on an ITO substrate by using a microwave-assisted liquid phase deposition method [22,23, 24]. In typical process, the $TiO_2$ nanorices were prepared by immersing a cleaned ITO substrate (sheet resistance ca. 9–22 $\Omega/cm^2$ purchased from VinKarola instuments USA) which was previously cleaned via an ultrasonication for 30 min in acetone and ethanol into a 10 mL of aqueous solution containing equimolar (0.05 M) solution of ammonium hexafluoro titanate ($(NH_4)_2TiF_6$) (AHT) and hexamethylenetetramine (HMT). Both chemical reagents were purchased from Sigma-Aldrich, USA, and used directly without any purification process. To obtain an optimum porosity properties and high percentage of (001) lattice plane of anatase $TiO_2$, the HMT concentration was varied from 0.03 to 0.08 M. In this case, the $TiO_2$ precursor molarity was fixed at 0.05 M. The growth time was 5 min, meanwhile the microwave power used was 180 W. After a growth process, the sample was taken out from the solution and rinsed with copious amount of pure water and dried under a nitrogen gas flow. Finally, the substrate was annealed in air at 350 °C for an hour.

Field Emission Scanning Electron Microscope (FESEM) technique (ZeiSS SUPRA 55VP) was used for examining the surface morphology of the sample. Meanwhile, the crystallinity of the nanostructure was examined via a high resolution transmission electron microscopy (HRTEM) analysis using Ziess Libra 200FE HRTEM apparatus operating at 200 kV. The X-ray diffraction spectroscopy (BRUKER D8 Advance with CuK$\alpha$ radiation and scan step as low as 2°/min) and the UV/VIS spectrometer (Lambda 900 Perkin-Elmer) were used to confirm the structure and the phase, and the optical properties of the sample, respectively.

## 2.2 Fabrication of dye sensitized solar cell and characterization

DSSC with a structure of ITO|$TiO_2$:dye|electrolyte|Pt electrode was fabricated utilizing the $TiO_2$ nanorice as the photo active layer. Prior to the device fabrication, a $TiO_2$ nanostructures modified-ITO substrate was immersed into a 0.05 mM ethanolic solution of dye (N719, purchased from Sigma-Aldrich, USA) for 12 h. It was then gently rinsed with ethanol and dried using a flow of nitrogen gas. For simplicity, we called this structure as as photoanode. A counter electrode was prepared by depositing a platinum layer of approximately 150 nm thickness on glass substrate via a sputtering method. A DSSC was assembled by clamping a photoanode and a counter electrode together. An iodide/tridiode redox couple (Iodolyte AN-50, purchased from Solaronix Switzerland) was used as the electrolyte and injected into the space between the photoanode and Pt counter electrode. The active area of the DSSC device was controlled at 0.24 $cm^2$.

The photovoltaic responses (*I-V* and incident photon to current efficiency (IPCE)) of the DSSC device was evaluated using a Keithley high-voltage source-measure unit (SMU) model 237 under AM 1.5 simulated irradiation (100 $mW/cm^2$) provided by 150 W Newport low-cost solar simulator. The photovoltaic properties of the DSSC device was characterized via an electrochemical impedance spectroscopy method using Solartron 1260 under a frequency range of 0.01 to 1 MHz, bias voltage at 0.5 V, and alternating current amplitude of 50 mA. The current amplitude is required to be higher in this case to accelerate the response and to avoid electrolyte drying during the measurement. However, our device is stable at this high-current source.

The performance of the samples in the DSSC device was verified at least for five times and the uncorrected standard deviation of the measurement was used to validate the performance of the device.

# 3 RESULTS AND DISCUSSION

## 3.1 $TiO_2$ nanorice characterization

$TiO_2$ nanorices have been successfully grown directly on an ITO substrate using the present approach after following a growth process for 5 min in a growth solution containing ammonium hexafluoro titanate (AHT) and hexamethylenetetramine (HMT). In the typical process, $TiO_2$ nanorice with length-scale from 6 to 70 nm were formed on the surface of ITO with a thickness can be up to 6 $\mu$m. Figure 1A shows typical FESEM image of the $TiO_2$ nanorice prepared using equimolar, i.e. 0.05 M, solution of AHT and HMT. As Figure 1A shows, high-density networked- $TiO_2$ nanorice forms on the surface covering the entire area of the substrate. Such networked-nanostructure produces a $TiO_2$ nanostructure films with a highly porous property (see inset in Fig. 1A), which is potential for solar cell application due to facilitating a high-dye loading and a facile diffusion of redox species on the surface of the nanostructure. As have been mentioned earlier, the networked- nanostructure is composed of nanoparticles with morphology resembles the rice shape. Probably due to a process of surface energy minimalization, they are connected each other, reflecting the individual $TiO_2$ nanorice bounded by a high-energy lattice plane. Figure 1B shows the corresponding X-ray diffraction spectrum of the samples. By comparing with the standard powder diffraction file for anatase $TiO_2$ (JCPDS file no. 21-1272), the obtained result is confirmed to be an anatase polymorph of $TiO_2$. By comparing with the JCPDS file, the $TiO_2$ nanorice's XRD spectrum peaks can be labeled as (101), (004), (200), (105) and (211) for peaks at $2\vartheta$ of 25.5, 37.8, 48.2, 54.0 and 55.0°, respectively. One important fact that can be noted from the result is the peaks ratio between (004) and (101) is quite high, i.e. 0.7, which is much higher compared to normal anatase nanostructure (approximately ranging from 0.2 to 0.4). This reflects that the anatase $TiO_2$ nanorice is characterized by dominant (001) lattice plane, the second highest in the surface energy. Thus, we expect that enhanced performance in applications, such as solar cell and photocatalysis, can be obtained from this new $TiO_2$ nanostructure.

The TEM analysis of $TiO_2$ nanorice structure is presented in Figure 2. A low resolution TEM image shown in Figure 2A verifies the morphology attained by FESEM results. A high resolution TEM analysis highlights the defect-less, smooth, and twinning-less lattice fringes with a spacing approximately 0.235 nm (see Fig. 2B), which reveals that the nanorice are single crystalline in nature, with their unidirectional growth on ITO substrates. This fringe spacing is corresponding to the facet (001), which is in a good agreement with the XRD results. The selected area electron diffraction (SAED) analysis of the nanorice (see inset in Fig. 2A) suggests an overlapping of two $TiO_2$ nanorice structures, which is depicted by two sets of bright spots, one of them is with high brightness (indexed diffraction pattern) and another with low brightness (pointed with arrows). The brighter set seems to be correspondent to the crystal at the top. The dimmer set can be correspondent to the crystal of $TiO_2$ nanorice placed at the bottom, as the image is due to diffraction of low energy scattered electrons or may be due to deviation from exact Bragg conditions, such as tilting of crystal (1–3°) and excitation errors. Nevertheless, it confirms that the nanorice is characterized by (001) high-energy lattice plane. A large exposure of high energy facet (001) can play a prominent role in the applications involving photolysis, catalysis and solar cells.

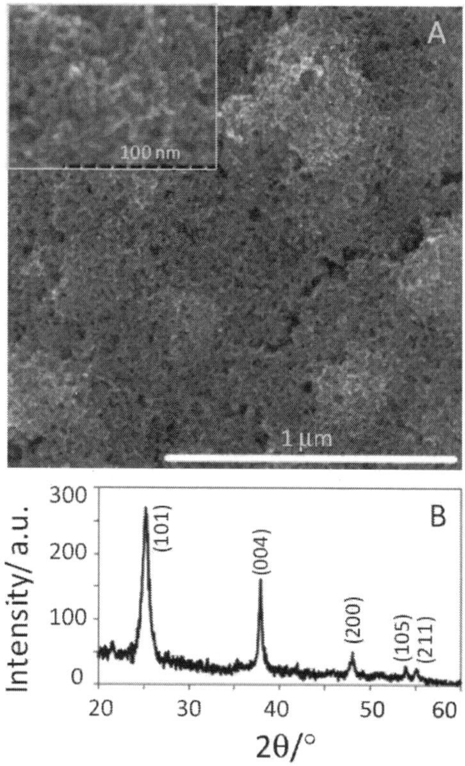


**Figure 1**. (A) Typical FESEM image of $TiO_2$ nanorice grown on an ITO substrate prepared using a growth solution containing 0.05 M of ammonium hexafluoro titanate (AHT), and 0.08 M HMT. Inset shows detailed structure of nanorice. (B) XRD pattern of the $TiO_2$ nanorice on ITO substrate.

Under normal liquid-phase deposition method, which uses AHT and boric acid as the growth solution, continuous films of $TiO_2$is obtained. In the present approach, while microwave energy applications only play a limited role in modifying nanocrystal growth morphology in the case of ZnO nanostructures [24, 25], in good agreement with the reported result by Parmar et al. [20] the microwave induces an anisotropic crystal growth in $TiO_2$, particularly nanorice-like morphology. By comparing the results obtained by them, which used acetylacetonate to decouple the hydrolysis and polycondensation of Ti ions with the result presented in this work, we remarked that the microwave energy likely induces an anisotropic stress and strain in the nanocrystallite and promotes the formation of anisotropic nanorice of $TiO_2$. And in the presence of the surfactant (HMT) here via an effective adhesion of its active amine functional onto the $TiO_2$ nanocrystallite, presumably on (001) plane, the nanorice of anatase $TiO_2$ that is bounded by this plane is realized. Such feature has also been found in the one-dimensional crystal growth in the case of ZnO [26, 27] and platelet [28, 29], brick-shape [30], nanofibrous [31, 32] and multipods [33] nanocrystal in the case of Au, Pt and Pd. Therefore, in order to obtain the extent role of HMT in the formation of (001)-faceted $TiO_2$ nanorice, we examined the nanocrystal growth properties under a different HMT concentration. The results are shown in Figure 3. As can be seen from the FESEM results, the length and the diameter of the nanorice decrease with the increasing of HMT concentration. This result reveals that the nature of nanorice packing and density can be controlled on the surface, which the density is increasing with the decreasing of nanorice dimension. Interestingly, from the figure, it was found that, although there is the change in the nanorice dimension, however, the aspect ratio; the length to diameter ratio, was relatively unchanged, namely 2.5. It was also observed that the morphology of the nanorice is unchanged with the variation in the HMT concentration. Despite no morphological modification, however, the change in the dimension as well as the nature of nanorice assembly on the surface may have produced novel properties for enhanced-performance in solar cell application. Table 1 summarizes the dimension of the nanorice prepared from several HMT concentrations with AHT concentration was fixed at 0.05 M.

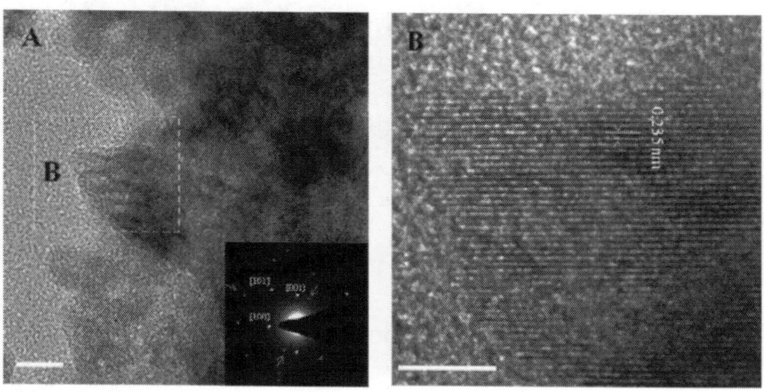


**Figure 2.** (A) Low resolution, and (B) high resolution transmission electron microscope images of TiO2 nanorice. The low resolution image verify nanorice morphology, and as well as SAED analysis (see inset in (A)), showing that nanorice are mono-crystalline. The high resolution image shows fringe spacing of 0.23 nm, reveals high exposure facet (001) with its growth along the [001] direction. Scale bars 10 nm in (A), and 5 nm in (B).

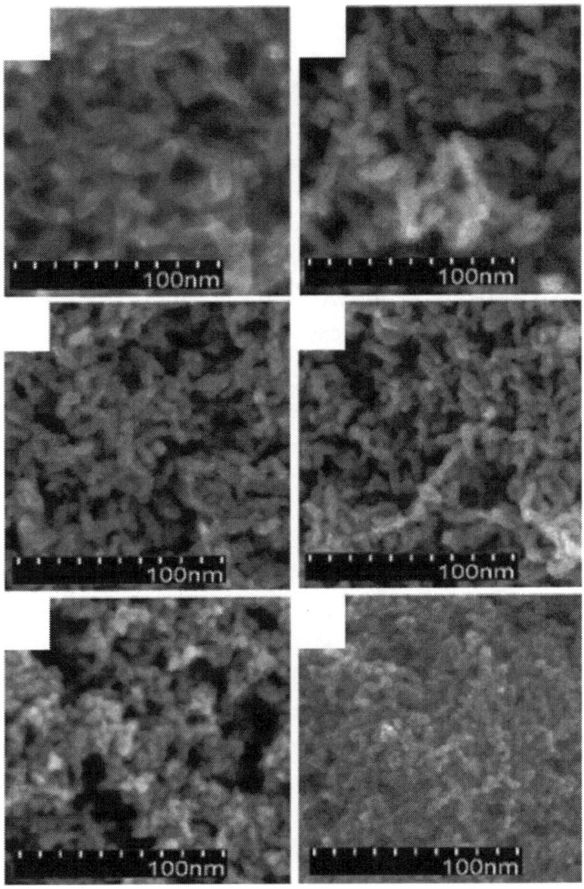


**Figure 3**. FESEM of $TiO_2$ nanorice structures grown on ITO substrate prepared using different HMT concentrations, namely 0.03 (A), 0.04 (B), 0.05 (C), 0.06 (D), 0.07 (E) and 0.08 M (F). $(NH_4)_2TiF_6$or (AHT) is fixed at 0.05 M. The growth time is 5 min.

While the morphology of the nanorice is relatively unchanged upon variation of HMT concentration, the crystalline properties of the samples were also evaluated by using the XRD analysis. The result is shown in Figure 4. As Figure 3 reveals, the crystallographic orientation, i.e. the lattice plane preference, is also found to be unchanged with the variation of HMT in the growth solution. Nevertheless, it was found the peaks intensity of X-ray diffraction from prominent lattice plane increases with the decreasing of nanorice dimension (HMT increasing), while, the full-width at half-maximum (FWHM) decreases with the decreasing of dimension. This reflects that the shrinking in the nanorice dimension might have improved the surface area of particular lattice plane. Thus, novel and enhanced properties are expected to be produced from the nanostructures. Figure5 shows the optical absorption spectra of the samples shown in Figure 3. In good agreement with the XRD results, the absorbance of the nanorice film effectively increases with the decreasing of the nanorice dimension. Judging from the FESEM results as shown in Figure 3, the increasing in the absorbance upon the decreasing in the nanorice dimension is resulted from the improvement of nanorice assembly, namely become more compact if the nanorice dimension reduces. A blue shift is also observed for the samples when the nanorice dimension reduced, which leads to the improvement of open-circuit voltage of the DSSC device [6]. Because of the nanorice assembly become more compact as the dimension reduced and considering the surface area of high-energy lattice plane increase, enhanced photoactivated surface reaction or charge-transfer [25] will be produced as the absorbance of the nanorice film increases with the decreasing of their dimension.

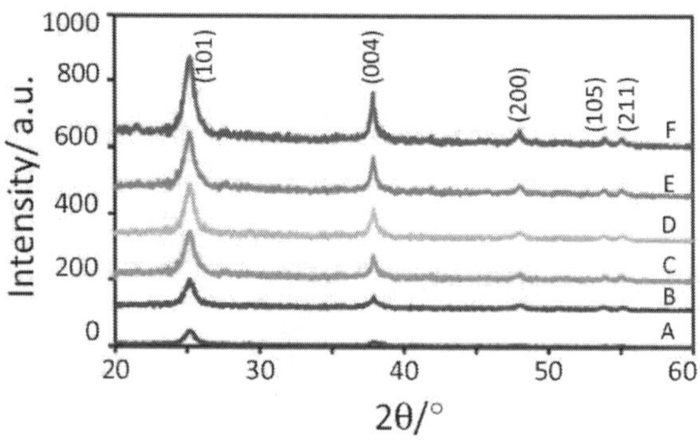


**Figure 4**. XRD spectra of $TiO_2$ nanorice prepared using different HMT concentrations, namely 0.03 (A), 0.04 (B), 0.05 (C), 0.06 (D), 0.07 (E) and 0.08 M (F). AHT was fixed at 0.05 M.

## 3.2 Solar cell characterization

A DSSC device with structure of ITO|$TiO_2$: dye (N719)|electrolyte ($I^{3-}/I^{2-}$)|Pt was fabricated to evaluate the photovoltaic property of the new structure. Figure 6 shows typical *J-V* curve for the DSSC device that were fabricated using six different $TiO_2$ nanorice structures of which their images are shown in Figure 3. As can be seen from Figure 5, the DSSC performance increases with the decreasing of the nanorice dimension, for example, the short-circuit current density ($J_{sc}$) and the open circuit voltage ($V_{oc}$) of the device enhanced from 1.58 ± 0.24mA/cm$^2$ and 0.56 ± 0.023V for the high grain size nanorice (device A) to 16.67 ± 0.265mA/cm$^2$ and 0.64 ± 0.00V for small grain nanorice (dimension) (device F). The increase in the performance of the device with the decreasing of nanorice grain size can be attributed to the high photon absorption by the device and possible enhanced electron transport [34, 35] as well as facile dye-$TiO_2$ charge transfer as the increase in the nanorice density and the surface are of high-energy (001) plane. The variation in the performance of the DSSC upon the variation of the nanorice size is unlikely related to the effect of surfactant because of the surfactant is seemed to be removed upon post-growth annealing at 350 °C for one hour. Therefore, it is clearly associated with the variation in the nanorice surface physico-chemistry. The photovoltaic parameters of the devices are summarized in Table 2.

**Table 1** $TiO_2$ nanorice prepared using different HMT concentrations with AHT fixed at 0.05 M.

| S. label | HMT(M) | Length (nm) | Width (nm) |
|---|---|---|---|
| A | 0.03 | 70 | 27 |
| B | 0.04 | 45 | 18 |
| C | 0.05 | 27 | 11 |
| D | 0.06 | 24 | 9 |
| E | 0.07 | 20 | 7 |
| F | 0.08 | 16 | 6 |

IPCE responses of the devices as shown in Figure 7 further verifies such phenomenon. The increasing value of IPCE of the device with the reducing of nanograin size stamped the role of high energy facet (001) exposure to generate photo electrons, and facilitates facile electron transportation. Hence, $J_{sc}$ is enhanced. As can be seen from Figure 5, the $V_{oc}$ of the device increases as the nanograin size decreases, reflecting the increasing of exciton lifetime or limited electron-hole recombination. Nevertheless, the device 'F' (device with the smallest nanograin size) shows a slight falls in $V_{oc}$, which can be attributed to the increasing of electron-hole recombination's rate [17, 36]. In spite of that fact, the device 'F' exhibits significant rise in $J_{sc}$, the result of enhanced exciton formation and facile electron transportation in the device probably due to greater exposure of high energy surface area.

**Table 2** Photovoltaic parameter of DSSCs device utilizing $TiO_2$ nanorice with nanograin size variation.

| Device | Conc. AHT: HMT (M) | $V_{oc}$ (V) | IPCE | $J_{sc}$ ($mA/cm^2$) | $R_{CT}$ ($\Omega$) | $R_S$ ($\Omega$) | $\eta$ (%) | $FF$ |
|---|---|---|---|---|---|---|---|---|
| A | 0.05:0.03 | 0.56 ± 0.023 | 9 | 1.58 ± 0.24 | 1014.52 | 42.39 | 0.32 ± 0.055 | 0.36 ± 0.004 |
| B | 0.05:0.04 | 0.62 ± 0.009 | 34 | 9.95 ± 0.18 | 587.15 | 42.46 | 2.41 ± 0.14 | 0.39 ± 0.004 |
| C | 0.05:0.05 | 0.64 ± 0.01 | 42 | 13.5 ± 0.22 | 405.36 | 40.62 | 3.24 ± 0.055 | 0.38 ± 0.004 |
| D | 0.05:0.06 | 0.68 ± 0.01 | 54 | 15.34 ± 0.27 | 307.74 | 38.9 | 3.73 ± 0.037 | 0.4 ± 0.004 |
| E | 0.05:0.07 | 0.68 ± 0.01 | 61 | 14.87 ± 0.20 | 268.81 | 33.88 | 3.81 ± 0.033 | 0.4 ± 0.006 |
| F | 0.05:0.08 | 0.64 ± 0.01 | 70 | 16.67 ± 0.265 | 224.27 | 34.76 | 4.05 ± 0.10 | 0.38 ± 0.004 |

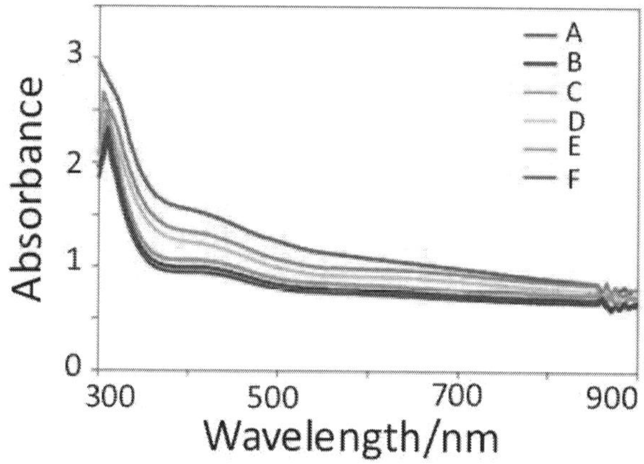


**Figure 5.** Typical optical absorption spectra of TiO2 nanorice prepare using different HMT concentrations.

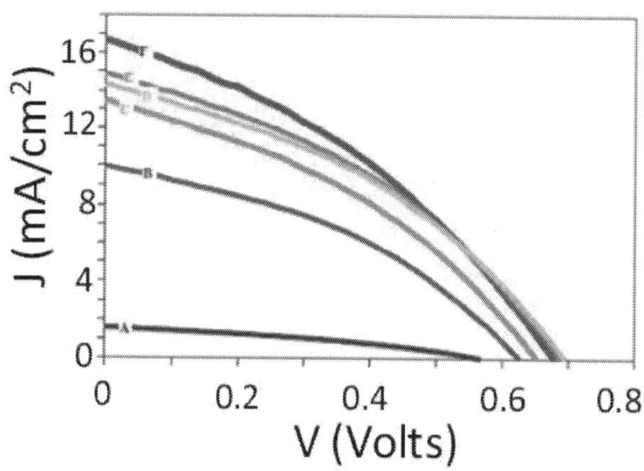


**Figure 6.** J-V characteristic of the DSSCs utilizing photoanodes with different nanograin size, namely bigger grain size (A) to smaller (F), under A.M1.5, 100 W illumination.

Electrochemical impedance spectroscopy (EIS) study of the devices (A-F) has explained such transportation of electrons and electron-hole recombination's natures in the device. The results are shown in Figure 8. It is observed that device 'A' with high$V_{oc}$ but low in fill factor, IPCE, and current density has higher value of charge transfer resistance ($R_{CT}$) in the region of Dye: $TiO_2$ |electrolyte interface, when compared with other devices 'B–F', rendering higher rate of recombination and weak charge transportation [2]. This is due to less reactive of (001) facet because of bulkier dimension. The $R_{CT}$ decreased to a great extent when the nanograin size reduce, as per expectation due to more (001) facet exposure. This may improve dye adsorption and interconnection between these nanoparticles [37]. Thus, the PCE increased.

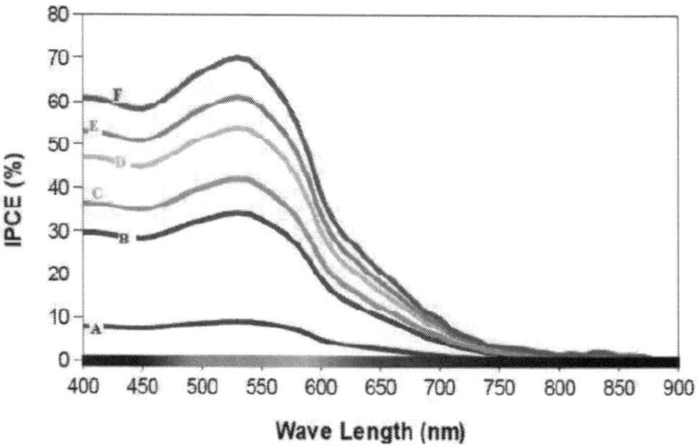


**Figure 7.** Incident photon to current efficiency of the devices under A.M1.5, 100 W illumination.

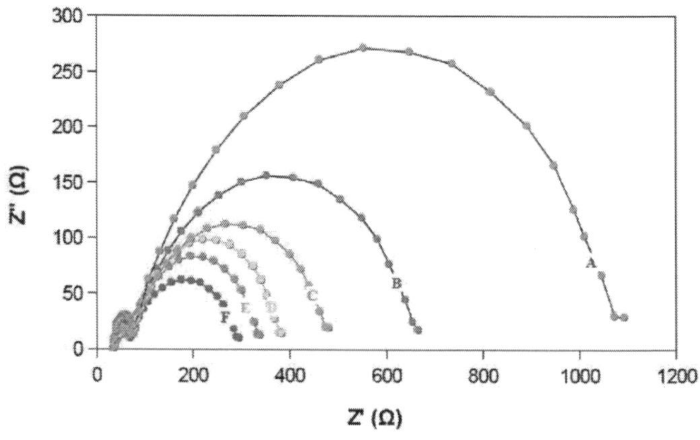


**Figure 8**. Electrochemical impedance spectra (EIS) of the devices.

# 4 CONCLUSIONS

Thin films of anatase $TiO_2$ with nanorice morphology and rich of (001) facet has been successfully synthesized directly on an ITO substrate surface via a liquid-phase deposition method using a growth solution containing hexamethylenetetramine (HMT) and ammonium hexafluoro titanate (AHT) under a microwave irradiation. The size of the nanorice as well as the basal plane of the nanocrystal can be finely controlled by varying the concentration of HMT in the reaction. It was found that the performance of the dye-sensitized solar cell was improved if the fraction of (001) facet in the nanostructures was increased. It was found that the performance of the DSSC device increases with the decreasing of nanorice dimension. The optimum device demonstrates the power conversion efficiency as high as 4.05 ± 0.10% with internal quantum efficiency (IPCE) as high as 72%. The decreasing in the nanorice dimension is predicted to increase the (001) facet exposure, enhancing the photoactivity, surface reactivity and electron transport in the device.

# ACKNOWLEDGMENTS

The authors would like to acknowledge the Ministry of Higher Education (MOHE), Malaysia for funding this work under research grants FRGS/1/2013/SG02/UKM/02/8 and HiCOE Project. The authors are also grateful for the financial support received from Ministry of Science, Technology and Innovation (MOSTI), Malaysia for the funding under Science Fund Grant (06-01-02-SF1157).

# REFERENCES

1. M. Graetzel, R.A.J. Janssen, D.B. Mitzi, E.H. Sargent, Nature **488**, 304 (2012)

2. M. Gratzel, Nature **414**, 338 (2001)

3. A. Ali Umar, S. Nafisah, S.K. Md Saad, S. Tee Tan, A. Balouch, M. Mat Salleh, M. Oyama, Sol. Energy Mater. Sol. C**122**, 174 (2014)

4. A.L. Linsebigler, G. Lu, J.T. Yates, Chem. Rev. **95**, 735 (1995)

5. W.-J. Ong, L.-L. Tan, S.-P. Chai, S.-T. Yong, A.R. Mohamed, Nanoscale **6**, 1946 (2014)

6. H. Ariga, T. Taniike, H. Morikawa, M. Tada, B.K. Min, K. Watanabe, Y. Matsumoto, S. Ikeda, K. Saiki, Y. Iwasawa, J. Am. Chem. Soc. **131**, 14670 (2009)

7. K. Lee, D. Kim, P. Roy, I. Paramasivam, B.I. Birajdar, E. Spiecker, P. Schmuki, J. Am. Chem. Soc. **132**, 1478 (2010)

8. J.R. Jennings, A. Ghicov, L.M. Peter, P. Schmuki, A.B. Walker, J. Am. Chem. Soc. **130**, 13364 (2008)

9. A.J. Frank, N. Kopidakis, J. van de Lagemaat, Coordin. Chem. Rev **248**, 1165 (2004)

10. E.J.W. Crossland, N. Noel, V. Sivaram, T. Leijtens, J.A. Alexander-Webber, H.J. Snaith, Nature **495**, 215 (2013)

11. C. Ducati, Nature **495**, 180 (2013)

12. E.L. Crepaldi, G.J.D.A.A. Soler-Illia, D. Grosso, F. Cagnol, F. Ribot, C. Sanchez, J. Am. Chem. Soc. **125**, 9770 (2003)

13. H. Zhang, Y. Wang, P. Liu, Y. Han, X. Yao, J. Zou, H. Cheng, H. Zhao, ACS. Appl. Mater. Interfaces **3**, 2472 (2011)

14. W.-Q. Wu, B.-X. Lei, H.-S. Rao, Y.-F. Xu, Y.-F. Wang, C.-Y. Su, D.-B. Kuang, Sci. Rep.-Uk **3**, 1352 (2013)

15. N. Wu, J. Wang, D.N. Tafen, H. Wang, J.-G. Zheng, J.P. Lewis, X. Liu, S.S. Leonard, A. Manivannan, J. Am. Chem. Soc.**132**, 6679 (2010)

16. A. Staniszewski, S. Ardo, Y. Sun, F.N. Castellano, G.J. Meyer, J. Am. Chem. Soc. **130**, 11586 (2008)

17. W. Shao, F. Gu, L. Gai, C. Li, Chem. Commun. **47**, 5046 (2011)

18. L.M. Falicov, G.A. Somorjai, Proc. Natl. Acad. Sci. **82**, 2207 (1985)

19. C.-Y. Chiu, P.-J. Chung, K.-U. Lao, C.-W. Liao, M.H. Huang, J. Phys. Chem. C **116**, 23757 (2012)

20. K.P.S. Parmar, E. Ramasamy, J. Lee, J.S. Lee, Chem. Commun. **47**, 8572 (2011)

21. C.-W. Peng, T.-Y. Ke, L. Brohan, M. Richard-Plouet, J.-C. Huang, E. Puzenat, H.-T. Chiu, C.-Y. Lee, Chem. Mater. **20**, 2426 (2008)

22. A.A. Umar, M.Y.A. Rahman, S.K.M. Saad, M.M. Salleh, M. Oyama, Appl. Surf. Sci. **270**, 109 (2013)

23. F.K.M. Alosfur, M.H.H. Jumali, S. Radiman, N.J. Ridha, M.A. Yarmo, A.A. Umar, Nanoscale Res. Lett. **8**, 1 (2013)

24. N.J. Ridha, A.A. Umar, F. Alosfur, M.H.H. Jumali, M.M. Salleh, J. Nanosci. Nanotechnol. **13**, 2667 (2013)

25. S.T. Tan, A.A. Umar, M. Yahaya, C.C. Yap, M.M. Salleh, J. Phys.: Conf. Ser. **431**, 012001 (2013)

26. L. Vayssieres, Adv. Mater. **15**, 464 (2003)

27. S.T. Tan, A.A. Umar, M. Yahaya, M.M. Salleh, C.C. Yap, H.-Q. Nguyen, C.-F. Dee, E.Y. Chang, M. Oyama, Sci. Adv. Mater. **5**, 803 (2013)

28. A.A. Umar, M. Oyama, M.M. Salleh, B.Y. Majlis, Cryst. Growth Design **9**, 2835 (2009)

29. A. Ali Umar, M. Oyama, M. Mat Salleh, B. Yeop Majlis, Cryst. Growth Design **10**, 3694 (2010)

30. A.A. Umar, M. Oyama, Cryst. Growth Design **8**, 1808 (2008)

31. A. Balouch, A. Ali Umar, A.A. Shah, M. Mat Salleh, M. Oyama, ACS Appl. Mater. Interfaces **5**, 9843 (2013)

32. A. Balouch, A.A. Umar, S.T. Tan, S. Nafisah, S.K. Md Saad, M.M. Salleh, M. Oyama, RSC Adv. **3**, 19789 (2013)

33. A.A. Umar, M. Oyama, Cryst. Growth Design **9**, 1146 (2009)

34. Q. Meng, T. Wang, E. Liu, X. Ma, Q. Ge, J. Gong, Phys. Chem. Chem. Phys. **15**, 9549 (2013)

35. M. Harb, P. Sautet, P. Raybaud, J. Phys. Chem. C **115**, 19394 (2011)

36. Z. He, C. Zhong, X. Huang, W.Y. Wong, H. Wu, L. Chen, S. Su, Y. Cao, Adv. Mater. **23**, 4636 (2011)

37. K. Zhu, N.R. Neale, A. Miedaner, A.J. Frank, Nano Lett. **7**, 69 (2007)

# Chapter 2

## Indium-free multilayer semi-transparent electrodes for polymer solar cells

**Alexander T. Barrows [a], Rob Masters [b], Andrew J. Pearson [a,1], Cornelia Rodenburg [b], David G. Lidzey [a,*]**

[a] Department of Physics and Astronomy, The University of Sheffield, Hicks Building, Hounsfield Road, Sheffield S3 7RH, UK

[b] Department of Materials Science and Engineering, The University of Sheffield, Sir Robert Hadfield Building, Mappin Street, Sheffield S1 3JD, UK

## ABSTRACT

We have explored the fabrication of indium-free electrodes for use in a PCDTBT:$PC_{70}BM$ organic photovoltaic (OPV) device, and compare different multilayer electrodes as the device anode. Two oxide/metal/oxide structures were investigated that consisted of $MoO_3$/Ag/$MoO_3$ (MAM) and $TeO_2$/Ag/$MoO_3$ (TAM) multilayers. Using scanning electron microscopy measurements, we find that the electrode utilising a $TeO_2$ seed layer encouraged the growth of a more continuous silver layer at low film thickness relative to an $MoO_3$ seed layer, and thus combines enhanced optical transmission (by around 7%) with low sheet resistance (14 Ω/□). This enhanced optical transmittance results in an increased short-circuit current in photovoltaic cells, with TAM-based devices having a power conversion efficiency around 6% higher than those fabricated using a comparable MAM electrode.

## KEYWORDS

Organic solar cell; Transparent electrode; ITO free electrode

## 1. INTRODUCTION

In the decade to the end of 2014, world primary-energy consumption grew on average by 2.1% per year [1]. Against this background there is increasing concern about climate change, resource constraints and the security of energy supply, leading to great interest in renewable energy generation. Photovoltaics are expected to play a major role in this sector, and whilst silicon currently dominates the market [2], organic photovolatics (OPVs) are a promising route to lowering the cost of solar power. The solution processed nature of the photoactive layer allows for the possibility of high throughput production via roll-to-roll techniques such as spray coating or printing [3] and [4], and in addition OPVs can be fabricated on flexible, lightweight substrates such as PET [5], [6] and [7] or even metal foils [8] and [9].

Due to its combination of high transmittance and good conductivity, indium tin oxide (ITO) is currently the ubiquitous material used as the transparent front electrode in OPVs. There are, however, concerns about the cost and scarcity of indium [7] and [10] and the embodied energy of the ITO layer [11] and [12]. Furthermore, high quality ITO is not easily compatible with the cheap, flexible polymer substrates that facilitate roll-to-roll production processes, as the brittle-nature of ITO results in a significant increase in sheet resistance after repeated bending [13] and [14]. Furthermore, such polymer substrates cannot tolerate the high temperatures that are commonly used in the deposition of highly conductive ITO. This second limitation leads to the sheet resistance of ITO on PET being around five times higher than that of similar ITO layers on glass [5], [15], [16] and [17].

There has thus been significant interest in the development of alternative materials for use as transparent conductive electrodes in OPVs. Here, possible replacements include the highly conductive polymer PEDOT:PSS in combination with metal grids [18] and [19], carbon nanosheets [20], silver nanowires [21], [22] and [23] and oxide/metal/oxide stacks. In such oxide/metal/oxide multilayer electrodes, the initial oxide layer acts as a 'seed layer' for the subsequent growth of an ultrathin metal film. When used in an OPV, the metal film provides high lateral conductivity ensuring a suitably low series resistance, with the second oxide layer facilitating charge extraction or charge-blocking from the device [24] and [25]. In addition to their electronic functionality, such oxide layers also suppress reflection from the metal film and maximise optical transmittance [26], [27],[28] and [29].

In this paper, we explore two different oxide/metal/oxide stacks as the hole extracting electrode in an OPV device based on the polymer:fullerene blend PCDTBT:$PC_{70}BM$. Whilst $MoO_3$/Ag/$MoO_3$ (MAM) electrodes have previously been investigated for use in OPVs [30], [31], [32] and [33], $TeO_2$/Ag/$MoO_3$ (TAM) electrodes have not. One study which investigated a $TeO_2$/ Ag/PEDOT:PSS electrode [34] suggested that a tellurium dioxide seed layer should lead to an enhanced short circuit current (*JSC*) in comparison with an equivalent $MoO_3$/Ag/PEDOT:PSS structure, resulting from the larger (real-part) refractive index of $TeO_2$[27]. Whilst the devices fabricated using such electrodes achieved a promising performance, the expected enhancement in *JSC* resulting from the use of $TeO_2$ was not verified experimentally. Here we find that replacement of the $MoO_3$seed layer by $TeO_2$ leads to significantly higher transmittance at low silver thickness, attributable to the formation of a more continuous silver layer, as observed by scanning electron microscopy (SEM). When applied as the window electrode in PCDTBT:$PC_{70}BM$ polymer solar cells the improved transmittance of the TAM electrode results in an increased *JSC* in comparison to devices based on a MAM electrode.

## 2. EXPERIMENTAL

The TAM and MAM devices explored were deposited upon 1.1 mm thick quartz-glass substrates. As a control, identical devices consisting of a $MoO_3$ hole-extraction layer deposited on an ITO electrode were also fabricated. Both quartz-glass substrates and pre-patterned ITO films were purchased from Ossila Limited. To prepare the quartz-substrates, they were initially cleaned by sequential sonication in warm Helmanex solution, IPA and deionised water for 5 min before being dried with compressed nitrogen and transferred to a nitrogen glovebox connected to a thermal evaporator. All subsequent thin-film evaporations were performed at a base pressure of $<5\times10^{-6}$ mbar through a shadow mask. Evaporation of molybdenum oxide pellets (99.95%, Testbourne), tellurium dioxide powder (99.995% trace metals basis, Sigma-Aldrich) and silver shot (≥99.99% trace metals basis, Sigma-Aldrich) were performed at rates of 0.3, 0.3 and 5.0 $Ås^{-1}$respectively. Transmittance spectra were measured using a Horiba FluoroMax-4 and are referenced to air. Sheet resistance was measured using a 4 point probe. SEM images were recorded using an FEI Nova NanoSEM 450 scanning electron microscope. A solid-state backscatter detector was used with a 3 kV primary beam, 5 mm working distance and a −4 kV beam deceleration field applied. Atomic force microscopy measurements were obtained using a Veeco Dimension 3100 with tips from Budget Sensors (300G-Al). Data analysis was performed using the Gwyddion software package.

The active layer of the OPV device was based on a blend of the polymer PCDTBT with the fullerene acceptor $PC_{70}BM$. The PCDTBT solution was prepared by adding the dry polymer to chlorobenzene, after which it was stirred at 70 °C for 24 h before being added to $PC_{70}BM$ to produce a 1:4 blend of PCDTBT:$PC_{70}BM$ at a concentration of 20 mg/ml. The solution was then heated to 70 °C and stirred for a further 24 h to fully solubilise both materials. This solution was cooled to room-temperature, filtered with a 0.45 µm PTFE filter and spin-coated onto the anode at a spin speed between 600 and 2000 rpm, producing a layer having a thickness between 50 and 105 nm (as measured using a Dektak surface profilometer). A top cathode layer was then deposited onto the active layer by a sequential thermal evaporation of 6.5 nm calcium followed by 100 nm aluminium. In all cases, devices thus had a layer structure of glass/anode/ PCDTBT: $PC_{70}BM$/Ca/Al. Finished devices were encapsulated using a glass slide and a UV epoxy (Ossila Ltd.) that was cured under a UV lamp for 30 min before being removed from the glovebox for testing. OPV devices were tested using a Newport 92251A-1000 solar simulator (AM1.5 spectrum at an intensity of 1000 $W/m^2$) through a 0.025 $cm^2$ aperture mask used to define the illuminated area. The data presented represents an average over 16 devices from 4 separate substrates for each type of anode tested. We only report the efficiency of the best 50% of devices, thus excluding pixels containing a significant defect, whilst avoiding the use of a statistical method that is open to user bias.

The optical properties of electrodes and OPV devices were modelled using a freely available program based on the transfer matrix method [35]. The optical constants for ITO, Ag, Al and Ca were taken from the library associated with this model, whilst those for $MoO_3$, $TeO_2$ and the photoactive blend were determined using an M2000v J.A. Woollam Co. spectroscopic ellipsometer.

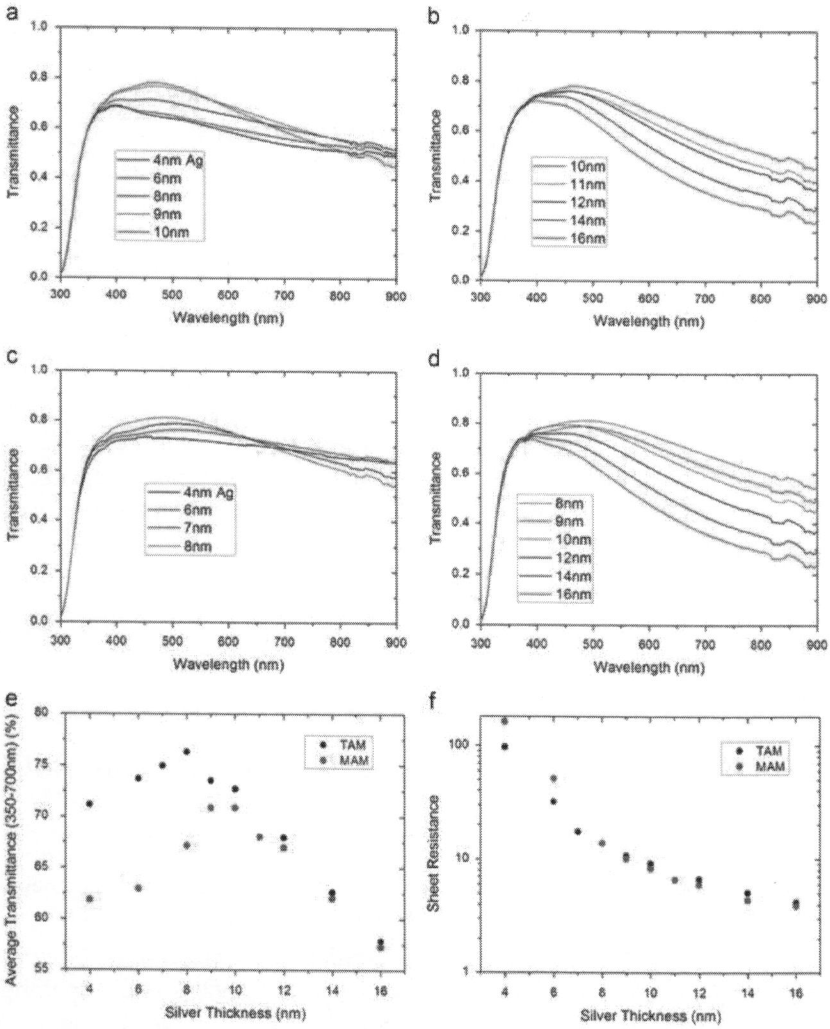


**Figure1.** Transmittance spectra for multilayer MAM ((a) and (b)) and TAM ((c) and (d)) electrodes on glass with varying thickness of the silver film. All spectra are referenced to air, meaning that the spectra presented herein include reflection and absorption by the glass substrate. Average transmittance in the 350–700 nm range (e) and sheet resistance (f) for varying silver thicknesses are also shown. In all cases the oxide layers are all kept at a thickness of 10 nm.

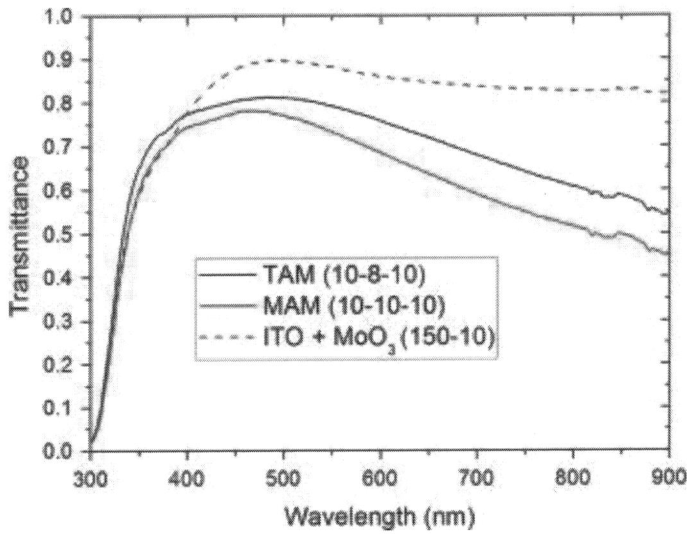


**Figure 2.** Transmittance spectra for the TAM and MAM multilayer electrodes with layer thicknesses of 10–8–10 nm and 10-10-10 nm respectively (i.e. with silver layer thickness corresponding to the percolation threshold for that structure). The spectrum for the ITO/$MoO_3$ anode used on reference devices is also presented.

# 3. RESULTS AND DISCUSSION

## 3.1. Optical and electrical properties of $TeO_2$ and $MoO_3$ based electrodes

To explore the effect of the structure of the oxide/metal/oxide multilayers on their optical properties, we fabricated a series of structures in which the thickness of the metal (silver) film was varied. The thickness of all films was determined by the film thickness monitor in the evaporation chamber. Note that such thicknesses should be considered as an average value, as the films are highly structured at nanometre length-scales (*vide infra*). In Fig. 1(a) and (b) we plot transmittance spectra for multilayer MAM electrodes as the silver-film thickness was varied between 4 and 10 nm (part (a)) and 10 and 16 nm (part (b)). In Fig. 1(c) and (d), we similarly show transmittance spectra for the TAM electrodes as the thickness of the silver was varied between 4 and 8 nm (part (c)) and 8 and 16 nm (part (d)). In all cases, the thickness of the oxide films (both $MoO_3$ and $TeO_2$) was fixed at

10 nm. In line with other reports on oxide/metal/oxide structures, we find that (counter-intuitively) the transmittance of the film initially increases with increasing thickness of the silver layer, before reaching a maximum at a thickness termed the percolation threshold. Above this thickness, the transmittance of the multilayer reduces, an effect particularly pronounced at longer wavelengths as a result of the silver film acting as a mirror. It can be seen that for our MAM electrodes the maximum transmittance occurs at a silver thickness of 10 nm; a result comparable to other literature values for the percolation threshold of Ag on $MoO_3$ (being between 10 and 11 nm [30], [31] and [32]). For the TAM electrodes, the maximum transmittance occurs at a reduced silver thickness of 8 nm. We summarise the average transmittance of the MAM and TAM electrodes over the wavelength-range 350–700 nm in Fig. 1(e), where it can be seen that over the Ag thickness range 4–8 nm average transmittance rises from 62–67% for the MAM structure and 71–76% for the TAM structure.

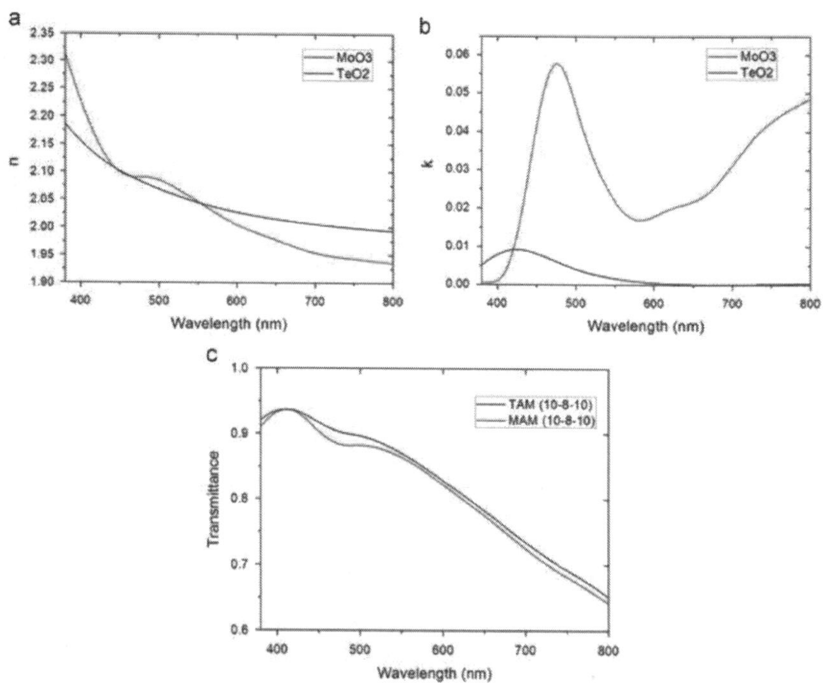


**Figure 3.** (a) and (b) show the optical constants for thin $MoO_3$ and $TeO_2$ layers, as determined by ellipsometry. Modelled transmittance spectra for $MoO_3/Ag/MoO_3$ (MAM) and $TeO_2/Ag/MoO_3$ (TAM) multilayer electrodes with layer thickness of 10–8–10 nm are shown in (c).

We have also measured the sheet resistance of the films as a function of the thickness of the silver layer, with data presented in Fig. 1f. It can be seen that for both the MAM and TAM electrodes there is a rapid decrease in sheet resistance as the thickness of the silver layer increases, however this decrease saturates as the film thickness exceeds the percolation threshold.

In Fig. 2, we illustrate the potential benefits of the use of the TAM structure as a transparent front electrode in an OPV, by plotting the relative transmission of the MAM and TAM electrodes at their percolation threshold (point of maximum optical transmittance). We find that the TAM electrode having an 8 nm thick silver layer has a maximum transmittance of 81.3%, an average transmittance (between 350 and 700 nm) of 76.3% and a sheet resistance of 13.9 Ω/□. In contrast a MAM electrode having a 10 nm thick silver layer has reduced maximum transmittance of 78.2% and an average transmittance (350–700 nm) of only 70.9%, with the sheet resistance being slightly lower at 8.3 Ω/□. For comparison, the reference ITO electrode (150 nm) coated with a 10 nm thick $MoO_3$ layer has a maximum transmittance of 89.8%, an average transmittance (350–700 nm) of 83.8% and a sheet resistance of 15.9 Ω/□. Figs. 1e and f clearly show that for low thickness of the silver film, the TAM electrode has a significantly higher transmittance and somewhat lower sheet resistance than the corresponding MAM electrode. As the thickness of the silver increases, the relative improvement in transmittance afforded by the use of a TAM structure is reduced, with the transmission of the TAM electrode being only slightly greater than that of the MAM electrode at and above the percolation threshold. Above an Ag thickness of 8 nm, the sheet resistances of the TAM and MAM electrodes are similar.

## 3.2. Origin of enhanced transmittance in $TeO_2$ based electrodes

The trends observed in transmittance and sheet resistance as the silver layer thickness increases result from the growth characteristics of the metal film. As silver is deposited, it initially forms isolated islands and chains (so-called Volmer–Weber growth), resulting in poor lateral conductivity due to a lack of continuous conduction pathways. As the film thickness increases, the silver particles coalesce to form an increasingly uniform and continuous layer. The formation of continuous conduction pathways then results in a rapid reduction in sheet resistance, a process that slows once a uniform and well-connected film has been created. Whilst a number of studies have observed these same trends for a variety of oxide/metal/oxide structures, there is some disagreement as to the exact source of the initial increase in transmittance as the layer is deposited. Some studies attribute it simply to light scattering by the unconnected islands, with this scattering effect being reduced as the layer becomes more uniform [36], [37] and [38]. Others attribute it to surface plasmon resonance at the optimum thickness, with excitation of the surface plasmon only being achieved effectively for the structure which obtains maximum transmittance (due to the different sizes and spacing of metallic chains in films with different thickness) [39], [40] and [41]. One recent study observed that as the film thickness increased, the film structure changed from one characterised by narrow cracks between metal islands and chains (usually unconnected) to a continuous structure with a random collection of nanoholes. It was proposed that the nanoholes allow for the formation of surface plasmon polaritons which propagate through the holes, with light then being re-radiated, thus increasing transmission. At higher thicknesses, such nanoholes become covered by the silver layer, and transmittance again drops [42].

We propose that our results can be explained by the formation of a more continuous silver layer at low thickness on a tellurium dioxide seed layer in comparison to silver films grown on a molybdenum oxide seed layer. The formation of a more continuous Ag layer on $TeO_2$ is consistent with the observed improvement in transmittance due to reduced scatter from unconnected islands, whilst the reduced sheet resistance at low silver thickness is similarly explained by the more continuous silver film. Overall, this effect allows the optimised TAM electrode to have a notably higher transmittance than can be obtained using a MAM electrode. This proposal is also consistent with the percolation threshold being reached at a comparatively reduced silver thickness on the $TeO_2$ seed layer compared to that observed on the $MoO_3$ seed layer (8 nm and 10 nm respectively). Clearly for Ag thicknesses above the percolation threshold for both types of electrode, the silver films formed are quasi-continuous and no large differences in transmittance are observed. The formation of a more continuous silver film on $TeO_2$ is commensurate with previous work in which liquid Ag was found to have a slightly lower contact angle on $TeO_2$than on $MoO_3$[43]. This is indicative of stronger interactions between the silver and the $TeO_2$, which will increase the energetic favourability of silver atoms becoming positioned on the metal oxide layer rather than aggregating into disconnected clusters. A lower contact angle will thus assist in the formation of a more continuous silver layer by encouraging growth to occur in a layer-by-layer manner (Frank–van der Merwe growth) and reducing the favourability of island formation. The lower contact angle on a $TeO_2$seed layer can be thought of as showing the improved wettability of liquid Ag on $TeO_2$, which enhances the uniformity of the silver film. We note that modification of the seed layer to achieve a higher surface energy (giving stronger substrate–silver interactions) has previously been found to lead to the formation of a more continuous silver film in MAM electrodes [44].

We have modelled the optical properties of the multilayer stacks using a Transfer-Matrix model to explore whether differences in the refractive index (real and imaginary) of the seed layers play a significant role in modifying the optical transmission of the electrodes[35]. The model assumes that all layers are uniform and continuous, and thus does not take into account the reduction in transmittance which occurs at silver film thickness below the percolation threshold for an oxide/metal/oxide structure. As input into this model, we have established $n$ and $k$ for $TeO_2$ and $MoO_3$ using spectroscopic ellipsometry as shown in Fig. 3(a) and (b) respectively. The calculated transmittance of MAM and TAM multilayer electrodes, (both having layer thickness of 10–8–10 nm) are shown in Fig. 3(c). It can be seen that our model suggests that over the wavelength region 380–800 nm, the MAM and TAM electrodes have an average transmittance of 81.2% and 82.1% respectively. The small enhancement in transmittance predicted for the TAM electrode results from the lower extinction coefficient of $TeO_2$ compared to $MoO_3$, as well as its favourable refractive index (higher at longer wavelengths, lower at shorter wavelengths – see Fig. 3(a)). This result strongly suggests that the enhancements in transmittance in the TAM structures observed for silver thickness less than 10 nm can be attributed to the formation of a more continuous silver film rather than any optical effects resulting from the differing seed layer materials themselves, with the slightly increased transmission for the TAM structures compared to the MAM structures for thicker Ag films attributable to the favourable optical properties of $TeO_2$.

To further explore the origin of the improved transmittance of the TAM electrodes at low Ag thickness, we have recorded SEM images of bilayer structures consisting of the metal oxide seed layer and an Ag layer. This is shown in Fig. 4, where we plot images for Ag films having a thickness of 6 nm and 8 nm on 10 nm thick seed layers of $TeO_2$ and $MoO_3$. It can be seen that the 6 nm thick silver film deposited on $MoO_3$ (part (a)) is characterised by a series of thin "chains" having a typical width in the range 10–30 nm. Such structures presumably result in a non-continuous pathway for charge-carrier extraction. In contrast, the 6 nm thick silver film deposited on $TeO_2$

(part (b)) appears to be more continuous in nature and consists of a series of holes having a diameter of around 5–20 nm. Importantly, such films contain fewer isolated (unconnected) features. When the silver thickness is increased to 8 nm, the film on $MoO_3$ (part (c)) appears to have formed a more connected structure, whilst the film on $TeO_2$ (part (d)) exhibits fewer holes than with the 6 nm thick film. In fact, when we compare the magnified images of an 8 nm Ag film on $MoO_3$ (part (c)) with a 6 nm Ag film on $TeO_2$ (part (b)) we find that they appear qualitatively similar. This finding is commensurate with our measurements of their optical properties, in which we found that the percolation threshold occurred for Ag films that were 2 nm thinner on a $TeO_2$ seed layer compared to those on $MoO_3$. Estimates of the coverage of the silver films from these SEM images also fit with this conclusion (seeTable S2).

## 3.3. Indium-free polymer solar cells

In order to demonstrate the practical use of such electrodes, we have fabricated bulk heterojunction polymer solar cells based on a PCDTBT:$PC_{70}BM$ photoactive layer using MAM, TAM and ITO/$MoO_3$ anodes. In each case, multilayer electrodes having the maximum transmittance were used; the MAM electrode thus utilised a 10 nm Ag layer whilst the TAM electrode incorporated a slightly thinner 8 nm Ag layer.

Again, we have used Transfer Matrix modelling to predict the optimum thickness of both the oxide layers in the multilayer electrode and the thickness of the PCDTBT:$PC_{70}BM$ active-layer [35]. The model allows us to predict the short-circuit current of a device by assuming an internal quantum efficiency of 100% (i.e. ignoring possible charge extraction losses). We term this photo-current as $J_{SC\_100\%}$. In our model, we adjust the thickness of the hole-extraction layer (the metal oxide layer placed adjacent to the active layer), the thickness of the seed layer positioned beneath the Ag and the thickness of the active semiconductor layer. We find that for seed layers having a thickness ≥30 nm, our model predicts that $J_{SC\_100\%}$ is maximised, however this photocurrent is only marginally larger than that predicted when using a 5 nm seed layer. In fact, we find that by adjusting the thickness of the active layer, the $J_{SC\_100\%}$ takes values between 11.50 and 11.59 mA/cm$^2$ as the seed layer thickness varies from 5 to 40 nm. We note, however, that thicker seed layers require the use of slightly thicker active layers (by around 10 nm) in order to maximise $J_{SC\_100\%}$. Experimentally, increasing the thickness of the active layer is known to increase loss within the device as a result of increased non-geminate charge recombination. In addition, we found that a thinner seed layer lead to a slight reduction in sheet resistance and a better correspondence to the modelled transmittance (see Figure S1 and Table S1). For this reason, we used a relatively thin (5 nm) seed layer for the MAM and TAM electrodes – an approach that has also been used in other studies [45]. With this seed layer thickness of 5 nm we find that the average transmittance in the 350–700 nm range rises slightly from 70.9% to 71.7% for the optimised MAM electrode and from 76.3% to 77.1% for the optimised TAM electrode. Our model predicted that for both of these electrodes the optimum value of $J_{SC\_100\%}$ would occur for a metal oxide HTL having a thickness of 10 nm with the active layer having a thickness of 70 nm. Experimentally, however, we found that the optimum device efficiency occurred when the active-layer was slightly thinner (60 nm) than predicted by the model for both types of multilayer electrode. For devices utilising an ITO/$MoO_3$ electrode, our model similarly predicted an optimum active layer thickness of 85 nm, however experimental measurements determined the highest device efficiency was realised for an active layer thickness 75 nm (see Tables S3, S4 and S5 for details). In both cases, the discrepancy between the results of the model and the experimental device optimisation likely results from imperfect charge extraction through the active layer; an effect not considered by the model.

We tabulate average device parameters for solar cells fabricated using the three different types of anodes in Table 1. We find that the optimised devices using a TAM electrode had an average PCE of (4.22±0.08)% whilst the optimised MAM-electrode devices had a slightly lower average PCE of (3.94±0.12)%. The improvement in efficiency for devices incorporating the TAM electrode results from a relative enhancement in *JSC* by 7% (8.70±0.12) mA/cm$^2$ compared with (8.11±0.14) A/cm$^2$; an effect that we attribute to the higher transmittance of the TAM electrode. The optimised ITO/$MoO_3$ reference device had the highest average PCE of (4.81±0.09)%, again resulting from an enhanced *JSC* of (9.12±0.24) mA/cm$^2$. It is worth noting that ITO/$MoO_3$ devices having an active layer thickness identical to that used in the TAM and MAM devices had a slightly lower PCE of 4.49%, with a comparable yet slightly lower *JSC* than the TAM devices (8.54±0.23) mA/cm$^2$ Current-density against voltage curves for typical solar cells incorporating each different anode are shown in Fig. 5.

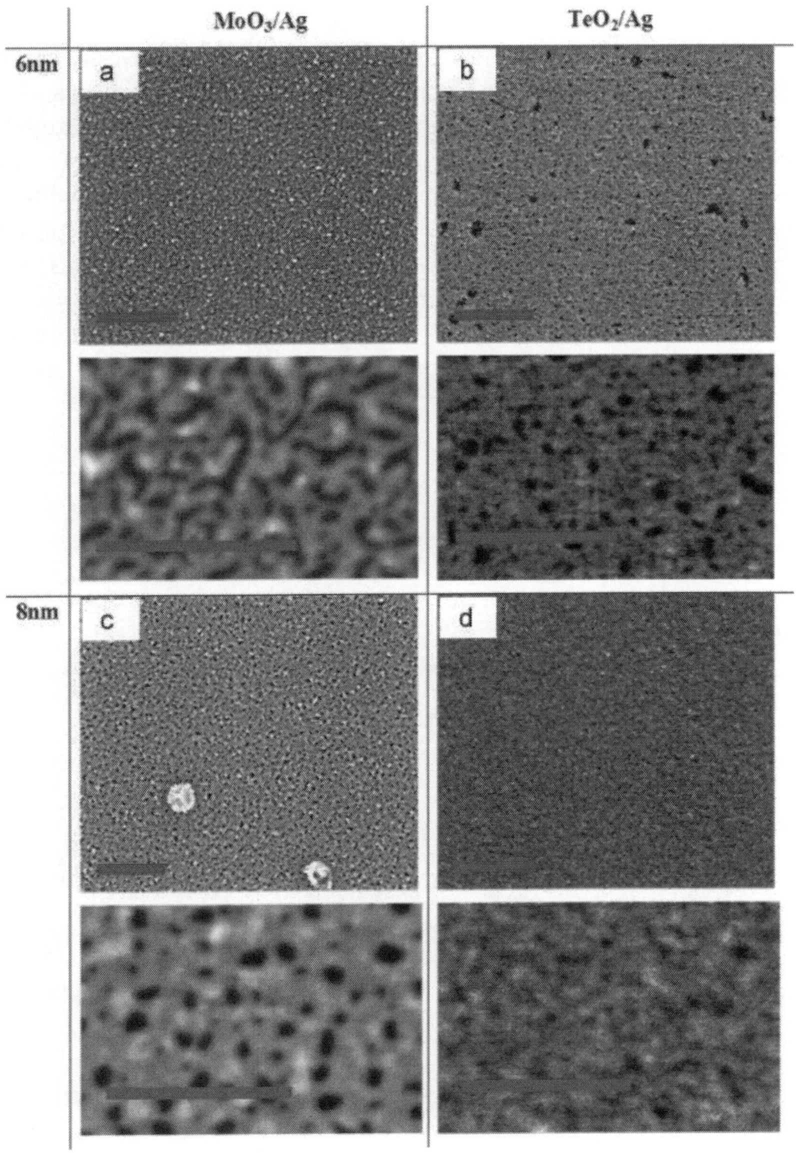


**Figure 4.** SEM images of $MoO_3$/Ag and $TeO_2$/Ag bilayers in which the oxide thickness is fixed at 10 nm. Parts (a) and (c) show $MoO_3$/Ag bilayer for a silver thickness 6 nm and 8 nm respectively. Parts (b) and (d) similarly show a $TeO_2$/Ag bilayer again utilising a silver layer having a thickness of 6 nm and 8 nm respectively. Blue scale bars correspond to 500 nm. In each case, we show a magnified image of the film surface in which the length of the (red) scale bar is set at 200 nm.

It is apparent that the ITO/$MoO_3$ based devices were also characterised by a larger *VOC*than were the devices utilising a MAM or TAM anode. This effect has been observed previously in MAM based devices [46] and [47] and has been attributed to diffusion of Ag through the charge extracting metal oxide layer, although such reductions in *VOC* are not always observed. It is not clear why such reductions in *VOC* are not uniformly reported; it is possible that, for example, variations in substrate temperature during the evaporation could result in differences in Ag diffusion. We note, however, that if such a reduction in*VOC* could be eliminated, then devices utilising the TAM electrode would outperform those based on ITO in which a comparable thickness of active layer has been used.

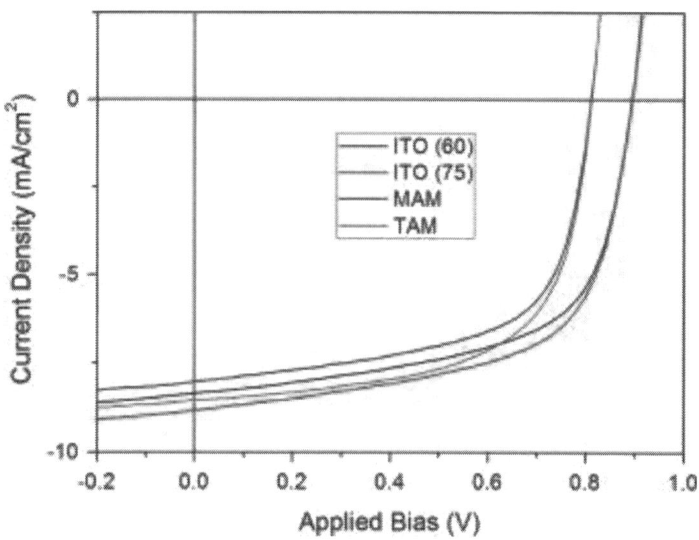


**Figure 5.** Current-density against voltage curves for devices fabricated using the different anodes.

**Table 1.** Electrode properties and solar cell performance for MAM (5–10–10), TAM (5–8–10) and ITO/$MoO_3$ (150–10) electrodes. The ITO electrode was tested with active layer thickness of both 60 nm and 75 nm. Standard deviations for device parameters are shown in brackets. Electrode properties are the mean (max/min) of three values.

| Anode | Electrode average transmittance 350–700 nm (%) | Electrode sheet resistance (Ω/□) | FF (%) | $V_{OC}$ (V) | $J_{SC}$ (mA/cm$^2$) | PCE (%) |
|---|---|---|---|---|---|---|
| $MoO_3$/Ag/$MoO_3$ (60 nm active) | 71.7(71.4/72.0) | 8.3(8.2/8.5) | 60.7 (1.1) | 0.801 (0.006) | 8.11(0.14) | 3.94 (0.12) |
| $TeO_2$/Ag/$MoO_3$ (60 nm active) | 77.1(76.9/77.4) | 13.9(13.7/14.0) | 60.5 (1.0) | 0.802 (0.007) | 8.70(0.12) | 4.22 (0.08) |
| ITO/$MoO_3$ (60 nm active) | 83.8(83.6/83.9) | 15.9(15.8/16.1) | 59.6 (0.7) | 0.882 (0.003) | 8.54(0.23) | 4.49 (0.16) |
| ITO/$MoO_3$ (75 nm active) | 83.8 (83.6/83.9) | 15.9(15.8/16.1) | 59.7 (0.7) | 0.885 (0.003) | 9.12(0.24) | 4.81 (0.09) |

# 4. CONCLUSION

In conclusion, we have fabricated $TeO_2$/Ag/$MoO_3$ (TAM) multilayer electrodes that have an average transmittance of 77.1% over the wavelength range of 350–700 nm. This transmittance is larger than the widely investigated multilayer system consisting of $MoO_3$/Ag/$MoO_3$ (MAM) that has an average transmittance of only 71.7% over the same wavelength range. When used as the semi-transparent electrode in a polymer solar cell, the improved transmittance of the TAM electrode results in an increased *JSC* in comparison to devices based on the MAM electrode (8.70 vs 8.11 mA/cm$^2$), leading to a small improvement in power conversion efficiency (4.22 vs 3.94%). We note however that such efficiencies are generally smaller than that of devices based on a conventional ITO electrode that reach an optimised efficiency of 4.81%. We conclude therefore that a TAM multilayer electrode appears to be a promising candidate as an indium free electrode for polymer solar cells, and that tellurium dioxide acts as an efficient seed layer in a multilayer electrode as it encourages the growth of a smooth and continuous silver film.

# ACKNOWLEDGEMENTS

AB acknowledges the EPSRC for funding via the E-Futures Doctoral Training Centre in Interdisciplinary Energy Research EP/G037477/1. We also acknowledge the UK EPSRC for supporting this research through grants EP/J017361/1 (Supersolar Solar Energy Hub) and EP/I028641/1 (Polymer / fullerene photovoltaic devices: new materials and innovative processes for high volume manufacture).

# REFERENCES

1. British Petroleum, BP Statistical Review of World Energy, 2015, ⟨www.bp.com/ statisticalreview⟩.

2. J. Nelson, C.J.M. Emmott, Can solar power deliver? Phil. Trans. R. Soc. A 371 (2013) 20120372.

3. T. Wang, N.W. Scarratt, H. Yi, A.D.F. Dunbar, A.J. Pearson, D.C. Watters, et al., Fabricating high performance, donor-acceptor copolymer solar cells by spraycoating in air, Adv. Energy Mater. 3 (2013) 505–512,

4. R.R. Søndergaard, M. Hösel, F.C. Krebs, Roll-to-roll fabrication of large area functional organic materials, J. Polym. Sci. Part B Polym. Phys. 51 (2013) 16–34,

5. Y. Galagan, I.G. de Vries, A.P. Langen, R. Andriessen, W.J.H. Verhees, S. C. Veenstra, et al., Technology development for roll-to-roll production of organic photovoltaics, Chem. Eng. Process. Process Intensif. 50 (2011) 454–461,

6. M. Kaltenbrunner, M.S. White, E.D. Głowacki, T. Sekitani, T. Someya, N. S. Sariciftci, et al., Ultrathin and lightweight organic solar cells with high flexibility, Nat. Commun. 3 (2012) 770,

7. F.C. Krebs, M. Jørgensen, K. Norrman, O. Hagemann, J. Alstrup, T.D. Nielsen, et al., A complete process for production of flexible large area polymer solar cells entirely using screen printing – first public demonstration, Sol. Energy Mater. Sol. Cells 93 (2009) 422–441,

8. Y. Galagan, D.J.D. Moet, D.C. Hermes, P.W.M. Blom, R. Andriessen, Large area ITO-free organic solar cells on steel substrate, Org. Electron. 13 (2012) 3310–3314,

9. F.-C. Chen, J.-L. Wu, W.-C. Huang, H.-M.P. Chen, W.-C. Chen, Flexible polymer photovoltaic devices prepared with inverted structures on metal foils, IEEE Electron Device Lett. 30 (2009) 727–729,

10. A.C. Tolcin, U.S. Geological Survey, 2010, Minerals Yearbook, Indium, n.d.

11. N. Espinosa, R. García-Valverde, A. Urbina, F.C. Krebs, A life cycle analysis of polymer solar cell modules prepared using roll-to-roll methods under ambient conditions, Sol. Energy Mater. Sol. Cells 95 (2011) 1293–1302,

12. A.L. Roes, E.A. Alsema, K. Blok, M.K. Patel, Ex-ante environmental and economic evaluation of polymer photovoltaics, Prog. Photovolt. Res. Appl. 17 (2009) 372–393,

13. R. Paetzold, K. Heuser, D. Henseler, S. Roeger, G. Wittmann, A. Winnacker, Performance of flexible polymeric light-emitting diodes under bending conditions, Appl. Phys. Lett. 82 (2003) 3342,

14. D.R. Cairns, R.P. Witte, D.K. Sparacin, S.M. Sachsman, D.C. Paine, G.P. Crawford, et al., Strain-dependent electrical resistance of tin-doped indium oxide on polymer substrates, Appl. Phys. Lett. 76 (2000) 1425,

15. E. Terzini, G. Nobile, T. Polichetti, P. Thilakan, Development and application of low temperature RF magnetron sputtered ITO thin films for a-Si:H based p/i/n junction solar cells, in: Proceedings of the 26th IEEE Photovoltaics Specialist Conference, 1997, pp. 667–670.

16. J.A. Thornton, High rate thick film growth, Annual Rev. Mater. Sci. 7 (1977) 239–260.

17. R. Bel Hadj Tahar, T. Ban, Y. Ohya, Y. Takahashi, Tin doped indium oxide thin films: electrical properties, J. Appl. Phys. 83 (1998) 2631,

18. W. Zhang, High-efficiency ITO-free polymer solar cells using highly conductive PEDOT: PSS/surfactant bilayer transparent anodes, Energy Environ. Sci. 6 (2013) 1956–1964,

19. Y. Galagan, B. Zimmermann, E.W.C. Coenen, M. Jørgensen, D.M. Tanenbaum, F. C. Krebs, et al., Current collecting grids for ITO-free solar cells, Adv. Energy Mater. 2 (2012) 103–110,

20. S.-I. Na, J.-S. Lee, Y.-J. Noh, T.-W. Kim, S.-S. Kim, H.-I. Joh, et al., Efficient ITOfree polymer solar cells with pitch-converted carbon nanosheets as novel solution-processable transparent electrodes, Sol. Energy Mater. Sol. Cells 115 (2013) 1–6,

21. J.-W. Lim, D.-Y. Cho, S.-I. Na, H.-K. Kim, Simple brush-painting of flexible and transparent Ag nanowire network electrodes as an alternative ITO anode for cost-efficient flexible organic solar cells, Sol. Energy Mater. Sol. Cells 107 (2012) 348–354,

22. K.-H. Choi, J. Kim, Y.-J. Noh, S.-I. Na, H.-K. Kim, Ag nanowire-embedded ITO films as a near-infrared transparent and flexible anode for flexible organic solar cells, Sol. Energy Mater. Sol. Cells 110 (2013) 147–153,

23. J.-H. Lee, H.-S. Shin, Y.-J. Noh, S.-I. Na, H.-K. Kim, Brush painting of transparent PEDOT/Ag nanowire/PEDOT multilayer electrodes for

flexible organic solar cells, Sol. Energy Mater. Sol. Cells 114 (2013) 15–23,

24. V. Shrotriya, G. Li, Y. Yao, C.-W. Chu, Y. Yang, Transition metal oxides as the buffer layer for polymer photovoltaic cells, Appl. Phys. Lett. 88 (2006) 073508,

25. R. Steim, F.R. Kogler, C.J. Brabec, Interface materials for organic solar cells, J. Mater. Chem. 20 (2010) 2499–2512,

26. J.C.C. Fan, Transparent heat-mirror films of TiO2/Ag/TiO2 for solar energy collection and radiation insulation, Appl. Phys. Lett. 25 (1974) 693, http://dx. doi.org/10.1063/1.1655364.

27. C. Guillén, J. Herrero, TCO/metal/TCO structures for energy and flexible electronics, Thin Solid Films 520 (2011) 1–17,

28. H. Lei, P. Qin, W. Ke, Y. Guo, X. Dai, Z. Chen, et al., Performance enhancement of polymer solar cells with high work function CuS modified ITO as anodes, Org. Electron. 22 (2015) 173–179,

29. F. Cheng, G. Fang, X. Fan, H. Huang, Q. Zheng, P. Qin, et al., Enhancing the performance of P3HT:ICBA based polymer solar cells using LiF as electron collecting buffer layer and UV-ozone treated MoO3 as hole collecting buffer layer, Sol. Energy Mater. Sol. Cells 110 (2013) 63–68,

30. F. Li, S. Ruan, Y. Xu, F. Meng, J. Wang, W. Chen, et al., Semitransparent inverted polymer solar cells using MoO3/Ag/WO3 as highly transparent anodes, Sol. Energy Mater. Sol. Cells 95 (2011) 877–880,

31. D.-T. Nguyen, S. Vedraine, L. Cattin, P. Torchio, M. Morsli, F. Flory, et al., Effect of the thickness of the MoO3 layers on optical properties of MoO3/Ag/MoO3 multilayer structures, J. Appl. Phys. 112 (2012) 063505,

32. L. Cattin, M. Morsli, F. Dahou, S.Y. Abe, A. Khelil, J.C. Bernède, Investigation of low resistance transparent MoO3/Ag/MoO3 multilayer and application as anode in organic solar cells, Thin Solid Films 518 (2010) 4560–4563,

33. N.P. Sergeant, A. Hadipour, B. Niesen, D. Cheyns, P. Heremans, P. Peumans, et al., Design of transparent anodes for resonant cavity

enhanced light harvesting in organic solar cells, Adv. Mater. 24 (2012) 728–732,

34. J.-F. Salinas, H.-L. Yip, C.-C. Chueh, C.-Z. Li, J.-L. Maldonado, A.K.-Y. Jen, Optical design of transparent thin metal electrodes to enhance in-coupling and trapping of light in flexible polymer solar cells, Adv. Mater. 24 (2012) 6362–6367,

35. G.F. Burkhard, E.T. Hoke, M.D. McGehee, Accounting for interference, scattering, and electrode absorption to make accurate internal quantum efficiency measurements in organic and other thin solar cells, Adv. Mater. 22 (2010) 3293–3297,

36. F. Li, Z. Lin, B. Zhang, Y. Zhang, C. Wu, T. Guo, Fabrication of flexible conductive graphene/Ag/Al-doped zinc oxide multilayer films for application in flexible organic light-emitting diodes, Org. Electron. 14 (2013) 2139–2143, http://dx. doi.org/10.1016/j.orgel.2013.05.023.

37. F. Nehm, S. Schubert, L. Müller-Meskamp, K. Leo, Observation of feature ripening inversion effect at the percolation threshold for the growth of thin silver films, Thin Solid Films 556 (2014) 381–384,

38. H.-K. Park, J.-W. Kang, S.-I. Na, D.-Y. Kim, H.-K. Kim, Characteristics of indiumfree GZO/Ag/GZO and AZO/Ag/AZO multilayer electrode grown by dual target DC sputtering at room temperature for low-cost organic photovoltaics, Sol. Energy Mater. Sol. Cells. 93 (2009) 1994–2002,

39. I.P. Lopéz, L. Cattin, D.-T. Nguyen, M. Morsli, J.C. Bernède, Dielectric/metal/ dielectric structures using copper as metal and MoO3 as dielectric for use as transparent electrode, Thin Solid Films 520 (2012) 6419–6423,

40. D. Zhang, H. Yabe, E. Akita, P. Wang, R. Murakami, X. Song, Effect of silver evolution on conductivity and transmittance of ZnO/Ag thin films, J. Appl. Phys. 109 (2011) 104318,

41. K.-H. Choi, H.-J. Nam, J.-A. Jeong, S.-W. Cho, H.-K. Kim, J.-W. Kang, et al., Highly flexible and transparent InZnSnOx/Ag/InZnSnOx multilayer electrode for flexible organic light emitting diodes, Appl. Phys. Lett. 92 (2008) 223302,

42. W.-F. Xu, C.-C. Chin, D.-W. Hung, P.-K. Wei, Transparent electrode for organic solar cells using multilayer structures with nanoporous silver

film, Sol. Energy Mater. Sol. Cells 118 (2013) 81–89, http://dx.doi.org/10.1016/j. solmat.2013.07.047.

43. D. Jeannot, J. Pinard, P. Ramoni, E.M. Jost, Physical and chemical properties of metal oxide additions to Ag-SnO2 contact materials and predictions of electrical performance, IEEE Trans. Compon. Packag. Manuf. Technol. 17 (1994) 17–23.

44. S. Schubert, J. Meiss, L. Müller-Meskamp, K. Leo, Improvement of transparent metal top electrodes for organic solar cells by introducing a high surface energy seed layer, Adv. Energy Mater. 3 (2013) 438–443,

45. Z. Wang, C. Zhang, R. Gao, D. Chen, S. Tang, J. Zhang, et al., Improvement of transparent silver thin film anodes for organic solar cells with a decreased percolation threshold of silver, Sol. Energy Mater. Sol. Cells 127 (2014) 193–200,

46. L. Cattin, Y. Lare, M. Makha, M. Fleury, F. Chandezon, T. Abachi, et al., Effect of the Ag deposition rate on the properties of conductive transparent MoO3/Ag/ MoO3 multilayers, Sol. Energy Mater. Sol. Cell. 117 (2013) 103–109,

47. T. Abachi, L. Cattin, G. Louarn, Y. Lare, A. Bou, M. Makha, et al., Highly flexible, conductive and transparent MoO3/Ag/MoO3 multilayer electrode for organic photovoltaic cells, Thin Solid Films 545 (2013) 438–444,

# Chapter 3

# Limit of efficiency for photon-enhanced thermionic emission vs. photovoltaic and thermal conversion

**Gideon Segev a, Yossi Rosenwaks a, Abraham Kribus b,**

a School of Electrical Engineering, Tel-Aviv University,Tel Aviv 69978, Israel

b School of Mechanical Engineering, Tel-Aviv University,Tel Aviv 69978, Israel

## ABSTRACT

Conversion of sunlight by photon-enhanced thermionic emission (PETE) combines a photonic process similar to photovoltaic cells, and a thermal process similar to conventional thermionic converters. As a result, the upper limit on the conversion efficiency of PETE devices is not the same as the Shockley–Queisser (SQ) limit that corresponds to the bandgap of the absorbing material, nor to the Carnot efficiency corresponding to its temperature. Here we analyze the upper limit on efficiency of ideal PETE devices in several possible configurations, in comparison to ideal photovoltaic cells and ideal solar thermal converters. Isothermal PETE converters are shown to be restricted to less than the SQ limit, but non-isothermal devices can exceed this limit. The limit of efficiency increases with the flux concentration reaching for example 52% at concentration of 1000 suns. Spectral splitting leads to a modest increase in conversion efficiency to 56% at 1000 suns. Addition of a secondary thermal cycle increases the efficiency limit for all PETE configurations, up to 69.8% and 70.4% for the cases of isothermal PETE and a dual bandgap PETE system at 1000 suns.

## KEYWORDS

Shockley–Queisser limit; PETE; Thermionic emission; Spectral splitting; Multi-junction cells

## 1. INTRODUCTION

Conversion of solar radiation to electricity is obviously subject to the laws of thermodynamics, which impose an upper limit on the conversion efficiency. However, different conversion paths lead to different limits on efficiency. These limits are well known for the two major paths of photovoltaic conversion and thermal conversion. The path of photon-enhanced thermionic emission (PETE) combines processes of both thermal and photovoltaic nature, and therefore the upper limit on its efficiency, and its relation to thermal-only and photovoltaic-only limits, needs to be clarified.

The maximum conversion efficiency of photovoltaic conversion in single-junction solar cells is subject to the Shockley–Queisser (SQ) limit [1]. This limit accounts for the three most important and inevitable losses in single-junction solar cells: transmission of sub-band gap photons, thermalization of high-energy electrons, and radiative recombination. The radiative recombination under non-equilibrium conditions considering photon recycling within the cell is described with a simple expression, derived by comparing the radiative recombination under equilibrium conditions to the absorption of environmental photons. According to this limit, the maximum conversion efficiency for a single-junction cell with optimal bandgap under standard illumination (AM1.5 spectrum, no concentration) is about 34%. Real cells have achieved efficiency of about 30% with optimal selection of the bandgap, close to this theoretical limit. Concentrating the solar radiation increases the maximum possible efficiency, for example to 41% at concentration of 1000 suns.

Surpassing the SQ limit requires lifting the fundamental assumptions leading to this limit. The most common approach is reducing the thermalization loss through the use of more than one bandgap, leading to spectral splitting and separate conversion of different parts of the solar spectrum. For example, in multi-junction solar cells, different sub-cells are arranged in a series optical arrangement, such that a photon transmitted through one material with higher bandgap may be absorbed in a subsequent material with lower bandgap [2]. For example, using the same assumptions as the SQ analysis, the upper limit on efficiency of a dual junction cell under 1000 suns is raised to 54.4%. In intermediate band solar cells, the absorber band gap is divided into two separate gaps for the same purpose [3]. Multi-junction cells achieve in practice efficiencies of over 43%, exceeding the SQ limit for a single-junction cell.

Solar thermal conversion usually requires three processes: conversion of radiation to heat, heat to mechanical work, and then work to electricity. This conversion path involves two unavoidable losses: blackbody emission from the radiation receiver in the first process, and heat rejection to the environment dictated by the second law of thermodynamics in the second process. The third process can, in principle, be performed without loss. The concentration of sunlight is also assumed to be ideal with no optical losses. The upper limit on solar thermal conversion efficiency can then be expressed by the following simple expression [4]:

$$\eta_{th} = \left(1 - \frac{\sigma T_R^4}{P_{sun}}\right)\left(1 - \frac{T_{amb}}{T_R}\right) \tag{1}$$

$T_R$ and $T_{amb}$ are the temperatures of the radiation receiver and the ambient, respectively. $P_{sun}$ is the flux density of incident solar radiation, determined by the optical concentration of sunlight. For each given concentration, the receiver temperature $T_R$can be optimized to yield the maximum conversion efficiency. For example, under concentration of 1000 suns ($P_{sun} = 1000\ \mathrm{kW/m^2}$) and $T_{amb} = 300\ \mathrm{K}$, the optimal receiver temperature is 1107 K and the upper limit on efficiency is 66.7%.

Another variation of solar thermal conversion is direct conversion from heat to electricity, using a purely thermionic converter heated by concentrated solar radiation. This direct conversion from heat to electricity replaces the second and third processes and of the thermo-mechanical option. Ideal thermionic converters can reach as high as 90% of the corresponding Carnot efficiency [5], and therefore the overall efficiency of solar thermionic conversion is also limited by Eq. (1).

A PETE device consists of two electrodes separated by a vacuum gap as illustrated inFig. 1. The cathode is a semiconductor with a bandgap that is matched to the solar spectrum, and with low electron affinity at its electron emission surface. When illuminated with concentrated solar energy, photogeneration increases the conduction band electron population above the equilibrium level, and the emission energy barrier is reduced. As a result, more electrons are emitted from the cathode at lower cathode temperatures compared to conventional thermionic emitters [6]. Moreover, thermalization processes within the cathode increase its temperature, further increasing the emission current density. Hence, PETE devices utilize both photonic and thermal processes for energy conversion, and may not be subject to either the SQ limit or the thermal limit. Schwede et al. [6] show that the PETE efficiency can surpass the SQ limit, but the conditions under which this limit is surpassed, and what is the new efficiency limit relevant to PETE converters, were not examined.

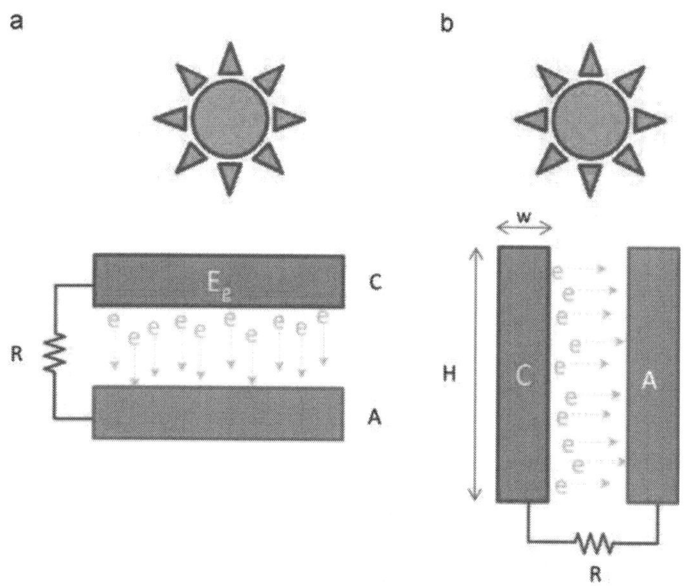


**Figure 1.** Schematic of PETE converters (a) front illuminated, (b) side illuminated.

Unlike traditional thermionic converters, PETE devices can convert solar energy even when both electrodes are at the same temperature [7]. Since the emitting surface is a selective contact for electrons, and the cathode contact is assumed to be perfectly selective for holes, isothermal PETE converters can be considered as equivalent to a Metal Insulator Semiconductor (MIS) solar cell [8]. In this case, power conversion is driven only by the photonic process without a temperature difference, and the efficiency should be constrained by the SQ limit.

Waste heat removed from a PETE converter can be used to generate additional electricity using a secondary converter, and increase the overall conversion efficiency. This includes heat from the anode in the front-illuminated case, Fig. 1(a) [6], and from both cathode and anode in the side-illuminated case, Fig. 1(b) [7]. This contribution can be significant since the anode temperature can be kept relatively high without affecting too much the performance of the thermionic part. Exploiting waste heat is also possible for photovoltaic converters, in particular concentrating photovoltaic (CPV) systems, but the heat is available at fairly low temperature since elevating the temperature will deteriorate the efficiency of the cells [9]. Therefore a secondary thermal stage to generate additional electricity is not very attractive. In the case of solar thermal conversion where the ideal Carnot cycle employs the ambient as the cold reservoir there is no option of exploiting the rejected waste heat.

A thermodynamic analysis of PETE converters was presented [10], where it was shown that the upper efficiency limit of a PETE device converges to the efficiency of solar thermal converters. However, that work did not consider the possibility of an IR coupling element that can absorb sub-bandgap photons and utilize their energy to further increase the cathode temperature and the conversion efficiency. Meir et al. compared the efficiency of ideal thermionic generators and PETE converters [11]. However, in their work, important aspects of the PETE operation such as the charge carriers' chemical potential were only estimated and not computed explicitly. Hence, these reported efficiencies cannot be considered as fundamental limits.

In this paper we analyze the limit on efficiency of ideal PETE converters, and compare to the SQ limit and to the ideal thermal conversion limit. Three PETE converter configurations are considered: isothermal, non-isothermal, and dual bandgap with spectral splitting. We show that under some conditions the maximum PETE efficiency is exactly the SQ limit, while under certain other operating conditions the PETE converter efficiency can exceed either the SQ limit or the ideal thermal conversion limit.

# 2. ANALYSIS

## 2.1. Single band gap devices

In order to derive the upper limit on the efficiency of an ideal PETE converter, a set of assumptions is used where only unavoidable losses are accounted for, in analogy to the detailed balance of the SQ limit [1]. The cathode conduction band electron population is a result of a particle balance comparing charge carrier generation and unavoidable losses, following the same principles as [12]. The charge carriers' concentration is uniform across the cathode. Radiative recombination is accounted for as in [1], and other types of bulk and surface recombination are neglected. The effect of the negative space charge in the vacuum gap is not considered. Electrons emitted from the anode that reach the cathode contribute to the conduction band electron population. It is also assumed that all the cathode emitted electrons are absorbed by the anode and vice versa, where the emitting surfaces of both electrodes have the same area. The cathode contacts are assumed to be perfectly selective in the sense that only holes can enter or exit through the cathode contacts. The anode is perfectly reflective to radiation, so that photons are lost to blackbody emission and net radiative recombination losses only through the surface that faces the incident solar radiation. Under steady state conditions the net electron current density will then be the difference between the total optical generation and the recombination:

$$q \cdot W\left[\Phi_S\left(E > E_g\right) - R'\right] = H\left(J_{em} - J_{rev}\right) \tag{2}$$

where $q$ is the electron charge, $\Phi_s(E>E_g)$ is the flux density of solar photons with energy above the bandgap, $R'$ is the non-equilibrium recombination per unit surface area,*Jem* is the emission current density from the cathode, *Jrev* is the current density of reverse emission from the anode, *W* is the cathode radiation absorbing area, and *H* is the cathode electron emitting area. In front-illuminated PETE devices, Fig. 1(a), the radiation absorbing area and electron emitting areas are identical. However, these areas can be different in side illuminated devices as illustrated in Fig. 1(b).

The emission current density from the cathode follows the derivation suggested by[12] and [13]

$$J_{em} = qn\sqrt{\frac{qK_B T}{2\pi \cdot m_n}}\, e^{-\frac{\chi}{K_B T}}, \qquad qV \le \phi_C - \phi_A$$

$$J_{em} = qn\sqrt{\frac{qK_B T}{2\pi \cdot m_n}}\, e^{-\frac{\chi + (\phi_A + qV - \phi_C)}{K_B T}}, \quad qV > \phi_C - \phi_A \tag{3}$$

where $n$ is the conduction band electrons concentration, $m_n$ is the electron effective mass and $\chi$ is the electron affinity. We note that the emission current density is proportional to the electron concentration and can be written as $Jem = KPETE(V) \cdot n$.

The reverse emission current density follows the standard thermionic emission formulation [5]:

$$J_{rev}(V) = AT_A{}^2 e^{-\frac{\phi_C - qV}{K_B T_A}}, \quad qV \le \phi_C - \phi_A$$

$$J_{rev}(V) = AT_A{}^2 e^{-\frac{\phi_A}{K_B T_A}}, \quad qV > \phi_C - \phi_A \tag{4}$$

$A$=120 A/cm$^2$ K$^2$ is the Dushman–Richardson constant and $V$ is the operating voltage.$\varphi_C$ and $\varphi_A$ are the cathode and anode work functions, respectively. $TA$ is the anode temperature and $KB$ is Boltzmann's constant.

The radiative recombination rate is calculated in the detailed balance limit [1]. Under equilibrium, the rate of photons emitted from a unit area is [6] and [14]

$$R_0 = \frac{2\pi}{h^3 c^2} \int_{Eg}^{\infty} \frac{(h\nu)^2 d(h\nu)}{e^{\frac{h\nu}{K_B T}} - 1} \tag{5}$$

This rate is increased exponentially under non-equilibrium conditions [14]:

$$R = R_0 \exp\left(\frac{E_{fn} - E_{fp}}{K_B T}\right) = R_0 \frac{n \cdot p}{n_{eq}\, p_{eq}} \tag{6}$$

where $E_{fn}$ and $E_{fp}$ are the conduction band and valence band quasi-Fermi levels, respectively, and $n_{eq}$, $p_{eq}$ are the equilibrium charge carrier concentrations. Therefore, the, non-equilibrium recombination per unit surface area R` is

$$R` = \frac{R_0}{n_{eq} \cdot p_{eq}} \left( n \cdot p - n_{eq} \cdot p_{eq} \right) = K_R \cdot \left( n \cdot p - n_{eq} \cdot p_{eq} \right) \tag{7}$$

Inserting Eqs. (4), (3) and (7) into Eq. (2), and defining *S=H/W* to be the cathode aspect ratio (thickness to width), yields

$$\Phi \left( E > E_g \right) + S \cdot \frac{J_{rev} (V)}{q} = K_R \left( n \cdot p - n_{eq} \cdot p_{eq} \right) + S \cdot \frac{K_{PETE} (V)}{q} \cdot n \tag{8}$$

Assuming equal hole and electrons excess carrier concentration $\delta n = n - n_{eq} = p - p_{eq}$ and rearranging we obtain

$$K_R \, \delta n^2 - \left[ S \frac{K_{PETE}}{q} + K_R \left( n_{eq} + p_{eq} \right) \right] \delta n + \left( \Phi \left( E > E_g \right) + S \frac{J_{rev}}{q} - S \frac{K_{PETE}}{q} \cdot n_{eq} \right) = 0 \tag{9}$$

The cathode's electron concentration is calculated by solving Eq. (9) for the excess carriers' concentration δn. The dependence of *n* on operating voltage is expressed through the voltage dependence of $K_{PETE}$ and $J_{rev}$. The net current density for a given voltage can then be obtained from Eqs. (4) and (3). The PETE conversion efficiency is then

$$\eta = \frac{\left( J_{em} - J_{rev} \right) V_{mpp}}{P_{sun}} \tag{10}$$

where *Vmpp* is the maximum power point (MPP) voltage, $P_{sun}$ is the concentrated radiation flux density incident on the PETE converter. $J_{rev}$ and $J_{em}$ are evaluated at the MPP voltage.

When operating in isothermal conditions, the temperature of both electrodes can be controlled via a heat exchanger in contact with both electrodes [7]. Under non-isothermal conditions, the cathode is thermally isolated from the surroundings and heat is removed to the surroundings only from the anode. Capturing the sub-bandgap radiation in the cathode in order to increase its temperature is possible if an IR coupling element is included in the cathode. This element must absorb all the sub-bandgap radiation without interfering with the other functionalities of the cathode. The cathode temperature is then determined through the energy balance of the cathode:

$$P_{sun} + J_{rev}\,\phi_B = \sigma T^4 + P_{rad} + J_{em}\,\phi_B \tag{11}$$

where $\sigma T^4$ is the blackbody emission from the front surface of the cathode. $P_{rad}$, the non-equilibrium radiative recombination loss, is [14]

$$P_{rad} = \left[\exp\left(\frac{E_{fn} - E_{fp}}{K_B T}\right) - 1\right]\frac{2\pi}{h^3c^2}\int_{E_g}^{\infty}\frac{(h\nu)^3 d(h\nu)}{e^{\frac{h\nu}{K_B T}} - 1} \tag{12}$$

The energy of emitted electrons from both cathode and anode corresponds to the emission potential barrier$\varphi_B$:

$$\begin{aligned} \phi_B &= \phi_C, \quad qV \le \phi_C - \phi_A \\ \phi_B &= qV + \phi_C, \quad qV > \phi_C - \phi_A \end{aligned} \tag{13}$$

The PETE net current density and cathode temperature can be calculated under a given set of operating conditions by coupling Eqs. (2) and (11), and solving for the electron concentration and cathode temperature.

In comparison to the PETE converter, according to the SQ derivation the net current density for an ideal photovoltaic cell follows:

$$J = q\left[\Phi_s\left(E > E_g\right) - R'\right] \tag{14}$$

The generation term is exactly as in Eq. (2); the recombination term is as in Eq. (7); and the quasi-Fermi levels splitting is exactly the operating voltage.

## 2.2. Dual bandgap devices

Spectral splitting may increase the conversion efficiency of PETE devices, as it does in photovoltaics. Fig. 2(a) shows two PETE sub-devices with different bandgaps, which are optically in parallel, and electrically independent. Fig. 2(b) shows an alternative configuration where the anode of the high band gap sub-device is attached to the low bandgap cathode, and the sub-devices are optically and electrically in series, similar to tandem PV cells. The optical efficiency of this configuration should be quite low because of the high reflection at several semiconductor–vacuum interfaces that the incident radiation encounters. Hence, only the independent configuration illustrated in Fig. 2(a) is investigated here.

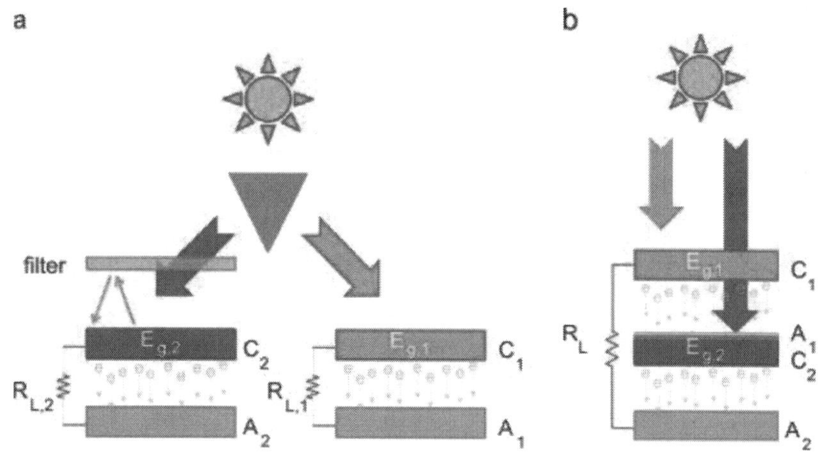


**Figure 2.** Schematics of dual band gap PETE converters, (a) optically in parallel, and (b) optically in series.

The spectral splitting step is ideal without optical losses. The higher band gap sub-device absorbs all the photons with energy above its band gap ($E>Eg_{,1}$), and the second sub-device absorbs all the remaining supra bandgap photons ($Eg_{,1}>E>Eg_{,2}$). A perfect filter that reflects all the photons with $E>Eg_{,1}$ is placed in front of the lower bandgap cathode[2], allowing recycling of high energy radiative recombination photons in the low bandgap cathode. With this filter, the lower band gap equilibrium radiative recombination term is reduced to

$$R_{0,2} = \frac{2\pi}{h^3c^2}\int_{Eg,2}^{Eg,1}\frac{(h\nu)^2 d(h\nu)}{e^{\frac{h\nu}{K_B T}}-1} \qquad (15)$$

Subscripts 1, 2 denote the higher and lower bandgap cathodes, respectively. Eq. (2) is solved for each device separately, where the optical generation is set according to the portion of the solar spectrum incident on each cathode.

The cathodes temperatures are again determined by an energy balance. However, in the dual bandgap case there can be several possible thermal configurations. We analyze the two extreme cases: (1) the two cathodes are thermally coupled with a common temperature that is determined by a common thermal balance; and (2) the two cathodes are thermally isolated, and each cathode has its own separate energy balance. It is assumed that the optical system directs all the IR radiation to the low bandgap cathode, which contains an IR coupling element. In the case (1) with both cathodes at the same temperature, the thermal balance is

$$P_{sun} + J_{rev,1}\,\phi_{B,1} + J_{rev,2}\,\phi_{B,2}$$
$$= P_{0,1} + P_{0,2} + P_{rad,1} + P_{rad,2} + J_{em,1}\,\phi_{B,1} + J_{em,2}\,\phi_{B,2} \qquad (16)$$

$P_{0,1}$ is the equilibrium radiative recombination power flux density from the high band gap cathode:

$$P_{0,1} = \frac{2\pi}{h^3c^2}\int_{Eg,1}^{\infty}\frac{(h\nu)^3 d(h\nu)}{e^{\frac{h\nu}{K_B T}}-1} \qquad (17)$$

$P_{0,2}$ is the equilibrium radiative recombination power flux density from the low band gap cathode:

$$P_{0,2} = \frac{2\pi}{h^3c^2}\int_{0}^{Eg,1}\frac{(h\nu)^3 d(h\nu)}{e^{\frac{h\nu}{K_B T}}-1} \qquad (18)$$

Note that the sum of $P_{0,1}$ and $P_{0,2}$ is the full-spectrum blackbody emission at temperature $T$ . The low band gap non-equilibrium radiative recombination loss, $P_{rad,2}$, follows:

$$P_{rad,2} = \left[\exp\left(\frac{E_{fn,2} - E_{fp,2}}{K_B\,T}\right) - 1\right]\frac{2\pi}{h^3c^2}\int_{Eg,2}^{Eg,1}\frac{(h\nu)^3 d(h\nu)}{e^{\frac{h\nu}{K_B T}}-1} \qquad (19)$$

The corresponding loss from the high band gap cathode is the same as Eq. (12).

In case (2) where the cathodes are thermally decoupled, a separate energy balance must be constructed for the cathode of each sub-device:

$$P_{sun,1} + J_{rev,1}\,\phi_{B,1} = P_{0,1} + P_{rad,1} + J_{em,1}\,\phi_{B,1}$$
$$P_{sun,2} + J_{rev,2}\,\phi_{B,2} = P_{0,2} + P_{rad,2} + J_{em,2}\,\phi_{B,2} \tag{20}$$

In this case an IR coupling element is attached to the low band gap cathode such that the incident solar radiation on each cathode is

$$P_{sun,1} = P_{sun}\,(E > E_{g,1})$$
$$P_{sun,2} = P_{sun}\,(E_{g,1} > E) \tag{21}$$

The voltages in each sub-device can be varied independently, and the MPP of the entire device corresponds to a set of voltages that maximize the total electrical output. The efficiency for dual bandgap device is

$$\eta = \frac{\left(J_{em,1} - J_{rev,1}\right)V_{mpp,1} + \left(J_{em,2} - J_{rev,2}\right)V_{mpp,2}}{P_{sun}} \tag{22}$$

## 2.3. Secondary thermal converter

The heat generated within the PETE converter must be removed to the environment. However, it can be harnessed to generate additional electrical power by mounting the PETE converter on top of a secondary thermal generator. The secondary generator can be for example an externally heated heat engine (e.g., Stirling or Rankine cycle) or a thermoelectric generator. An ideal thermal converter is represented by a Carnot heat engine that has the highest possible efficiency for the given temperatures.

For a single bandgap non-isothermal PETE converter the heat available to the secondary converter originates from electron thermalization in the anode:

$$Q = \left(J_{em} - J_{rev}\right)\left(\phi_{B,C} - V\right) \tag{23}$$

For the dual band gap configuration the heat flux density to the thermal cycle is the sum of the contribution of the two sub-devices:

$$Q = \left(J_{em,1} - J_{rev,1}\right)\left(\phi_{B,C1} - V_1\right) + \left(J_{em,2} - J_{rev,2}\right)\left(\phi_{B,C2} - V_2\right) \tag{24}$$

In the isothermal case, heat can be extracted from both the cathode and anode. The heat flux density to the secondary converter can be obtained from the overall energy balance[7]:

$$Q = P_{sun} - P_{PETE} - P_{rad} - \sigma T^4 \tag{25}$$

Assuming that the secondary converter is an ideal heat engine operating at Carnot efficiency, the secondary cycle output will be

$$P_{th} = Q\left(1 - \frac{T_{amb}}{T_A}\right) \tag{26}$$

The secondary thermal cycle efficiency is

$$\eta_{sec} = \frac{P_{th}}{P_{sun}} \tag{27}$$

and the overall conversion efficiency then becomes

$$\eta_{total} = \eta_{PETE} + \eta_{sec} \tag{28}$$

# 3. RESULTS

## 3.1. Parameter definitions

A general and fundamental limit on conversion efficiency should depend on a minimum number of parameters that describe specific material properties and the operating conditions. For PV cells, the SQ derivation depends only on the concentration as an arbitrarily selected input. The cell temperature should be as low as possible, and can be fixed at ambient temperature. The cell contacts are assumed to be ideally selective, and their Fermi levels are exactly the semiconductor absorber's quasi-Fermi levels. Hence, there is no need to assign them any specific material properties. Two last parameters, the material bandgap and the operating voltage, can be used as free parameters for optimization for a given concentration. The ideal solar thermal converter also depends only on the concentration as an arbitrarily selected input. The cold reservoir temperature should be as low as possible, and can be fixed at ambient temperature. The receiver temperature is a free parameter for optimization, and for each concentration an optimal receiver temperature can be found.

The PETE efficiency depends on many more parameters that need to be defined. These include the operating conditions: flux concentration, cathode and anode temperatures, and operating voltage, as well as material properties: cathode bandgap, doping, and electron affinity, and anode work function. In previous analyses, the optimal bandgap was shown to be close to 1.4 eV [6], similar to PV cells. The efficiency decreases monotonically with cathode electron affinity, leading to a recommended value of 0 eV[12]; yet, at elevated temperatures, somewhat higher electron affinities can be used without a significant reduction in efficiency. For example, for temperatures above 500 K, electron affinities of 0 eV and 0.4 eV would produce the same efficiency [12]. The cathode should be highly doped p-type, and the anode work function should be as low as possible. Previous work [7] has shown that the cathode geometric aspect ratio can also influence the conversion efficiency. Hence, this parameter should be considered as well.

These trends found in past work can be used to fix some of the parameters, allowing investigation of the remaining free parameters. Fixing these values takes into account the range of properties available with real materials. The anode work function was set to 0.9 eV, the lowest value measured in practice [15]. Unless stated otherwise, the anode temperature was fixed at 500 K; the possibility to optimize the anode temperature is discussed in Section 4. The cathode is p-type doped with a concentration of $10^{19}$ $cm^{-3}$and band gap of 1.4 eV. The input power and photon flux density are calculated according to the AM1.5D spectrum multiplied by the solar flux concentration.

## 3.2. Single bandgap isothermal PETE

Fig. 3 shows the isothermal PETE efficiency as a function of the flux concentration for several values of *S*. The cathode band gap is optimized at every concentration for highest efficiency; the operating temperature is constant at 700 K. The cathode is $10^{19}$ $cm^{-3}$ p-type doped, its electron affinity is 0.4 eV, and the anode work function is 0.9 eV. Also shown is the SQ limit for the same temperature and concentration. The PETE efficiency is identical to this limit at low concentrations, but is below the limit as the concentration increases. This is a result of the increased radiative recombination: at low concentration, the excess carriers' concentration is low and the added recombination is negligible. However, since radiative recombination is proportional to the square of the electrons concentration, it becomes more significant at high concentration. Increasing the cathode aspect ratio on the other hand, reduces the electrons concentration and keeps the isothermal PETE efficiency close to the SQ limit for higher flux concentrations.

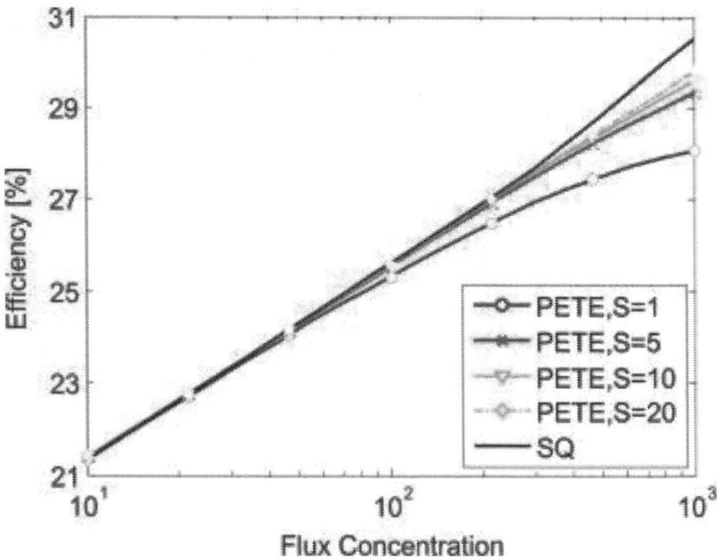


**Figure 3.** SQ efficiency and isothermal PETE efficiency vs. concentration for several cathode area ratios *S*. The solar cell and PETE electrodes temperature is 700 K. The cathode bandgap is optimized for every flux concentration. The cathode is $10^{19}$ $cm^{-3}$ p-type doped, the electron affinity is 0.4 eV, and anode work function is 0.9 eV.

The behavior shown in Fig. 3 can be better understood by considering in more detail the operation of the isothermal PETE converter. When the cathode and anode temperatures are equal, power generation is driven only by the charge carriers' non-equilibrium concentrations in the illuminated cathode. Fermi level splitting is then an essential condition for non-zero short circuit current density in isothermal PETE devices [7]. Since both electrodes are at the same temperature, the net current direction is determined by electron emission barriers in the cathode and the anode. We define $\Delta E$ to be the difference between the cathode conduction band quasi-Fermi level and the equilibrium Fermi level. Fig. 4 shows a schematic band diagram of a PETE device where the operating voltage is exactly the $\Delta E$. In this case, the emission barrier for both electrodes is identical and the net current will be zero. For any voltage below $\Delta E$ the emission barrier from the cathode is reduced, the net electron current is from the cathode to the anode, and power is produced. Hence, the maximum power point voltage must be below $\Delta E$.

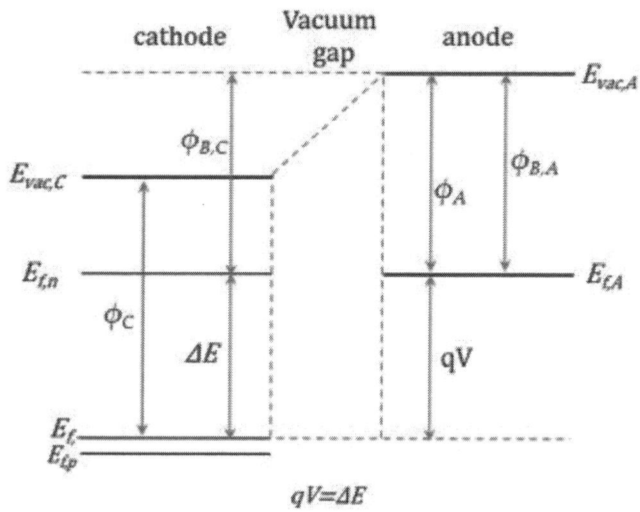


**Figure 4.** PETE schematic band diagram.

In contrast to PETE, for PV cells under the SQ assumptions the Fermi level splitting is exactly the operating voltage. Hence, radiative recombination, which is exponentially dependent on the difference between the quasi-Fermi levels, is more dominant in PETE devices. Fig. 5 shows the difference in quasi-Fermi levels as a function of the operating voltage for PETE cathodes with several aspect ratios and anode work functions. The cathode is $10^{19}$ $cm^{-3}$ p-type doped, the band gap is 1.373 eV, the electron affinity is

0.4 eV, the electrodes temperature is 700 K and the flux concentration is 500. $E_{f,n}-E_{f,p}=qV$ is shown as a diagonal line. The maximum power points are marked by stars. As can be seen in Fig. 5, the difference in quasi-Fermi levels is higher than the operating voltage in nearly all cases. As a result, radiative recombination is higher in PETE devices compared to an ideal PV cell at the same operating voltage, reducing the PETE efficiency below the SQ limit. When the operating voltage is below the difference in work functions, the electron concentration is constant and so is the difference in quasi-Fermi levels. For voltages above the difference in work functions, the electron concentration increases with voltage, leading to a corresponding increase in the difference of the quasi-Fermi levels and the radiative recombination [12]. Hence, in order to approach the SQ limit this increase should be at the highest voltage possible, and lower anode work functions are desirable.

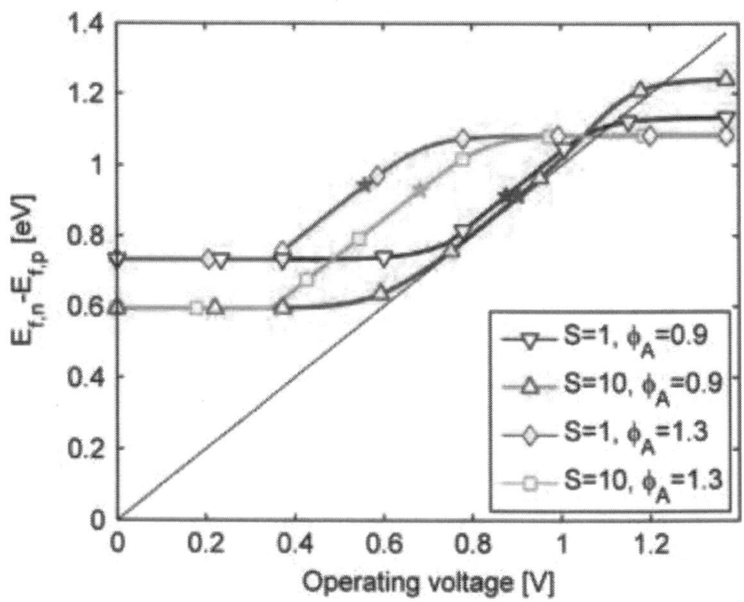


**Figure 5.** Difference in quasi-Fermi level in isothermal PETE as a function of the operating voltage for several anode work functions and electrodes aspect ratios. The cathode is $10^{19}$ $cm^{-3}$ p-type doped, the band gap is 1.373 eV, and the electron affinity is 0.4 eV. The electrodes temperature is 700 K and the flux concentration is 500. The maximum power points are marked by stars.

An increase in the aspect ratio *S* increases the cathode volume and reduces the cathode electrons concentration, and with it the difference in quasi-Fermi levels and the radiative recombination. However, this does not reduce the net current, because of the increased area for electron emission. Hence, when the aspect ratio is high enough, the quasi-Fermi levels splitting will coincide with the operating voltage at the MPP. In this limit, both isothermal PETE and the SQ calculation will have the same radiative recombination and efficiency. Note however that very large values of *S* may be impractical if the physical dimension exceeds the charge carriers' diffusion length.

Fig. 6(a) shows the total efficiency of a two-stage device with an isothermal PETE converter and a secondary thermal converter, as well as the separate efficiencies of each stage, as a function of the flux concentration and for electrodes aspect ratios of 1 and 20.Fig. 6(b) shows the corresponding optimal operating temperature. The cathode is $10^{19}$ $cm^{-3}$ p-type doped, the band gap is 1.4 eV, and the electron affinity is 0.4 eV. The anode work function is 0.9 eV. The total efficiency varies from 40% for low flux concentration and low aspect ratio, up to 69.8% for flux concentration of 1000. The photonic contribution is significant at low flux concentration, where the temperature is low and the thermal stage efficiency is low. In this range, increasing the cathode aspect ratio is effective, and the photonic contribution in the high aspect ratio case peaks at 27.9% at flux concentration of 32 and temperature of 527 K. For flux concentrations above 400 the aspect ratio has little effect on the conversion efficiency. This is due to operation in the thermionic-emission dominated regime [16] where electron concentration and recombination loss in the cathode are already low. The contribution of the secondary thermal cycle increases with operating temperature and becomes the dominant contribution at concentrations above 100.

Also shown in Fig. 6(a) is the ideal thermal cycle efficiency. The receiver temperature was chosen to maximize the efficiency at each concentration, and the heat sink temperature is 300 K. The efficiency of the two-stage isothermal PETE is considerably higher than the ideal thermal cycle efficiency over the entire range of flux concentration. This is due to the photonic contribution of the isothermal PETE device.

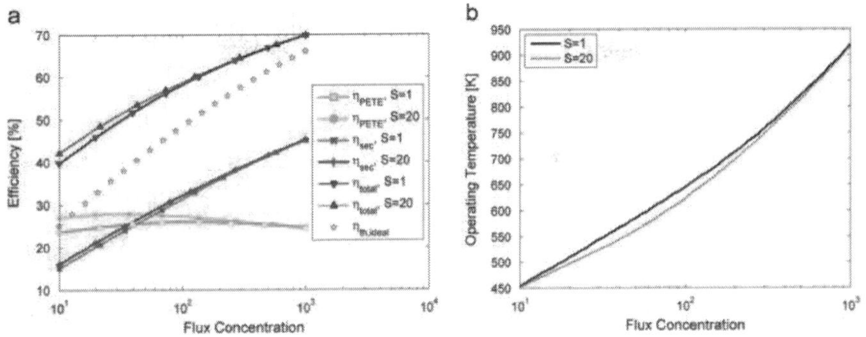


**Figure 6. (a)** Efficiency of the optimal isothermal PETE, the corresponding secondary thermal cycle, and total efficiency, and (b) the corresponding optimal operating temperature, both as a function of the flux concentration and for electrodes aspect ratios of 1 and 20. The cathode is $10^{19}$ $cm^{-3}$ p-type doped, the band gap is 1.4 eV, and the electron affinity is 0.4 eV. The anode work function is 0.9 eV.

## 3.3. Single bandgap non-isothermal PETE

As discussed above, radiative recombination keeps the isothermal PETE efficiency below the SQ limit. An added thermal contribution driven by a temperature difference may then surpass this limit. Fig. 7(a) shows the efficiency for non-isothermal PETE, where the cathode temperature is calculated from the energy balance at each flux concentration, for several cathode aspect ratios. The anode temperature is fixed at 500 K, the cathode is $10^{19}$ $cm^{-3}$ doped p-type, and the band gap is 1.4 eV. Fig. 7(b) shows the corresponding cathode temperature. Also shown in Fig. 7(a) is the SQ efficiency calculated for the same concentrations at temperatures of 300 K and 400 K, representing a range of practical operation temperatures for PV cells. It should be noted that comparing to the SQ efficiency where the cell is at the same temperature as the PETE cathode yields significantly lower efficiencies, for example 21% and 15% at flux concentrations of 100 and 1000, respectively. Therefore a comparison to practical PV cell temperatures is deemed more relevant.

As expected, for low flux concentrations and cathode temperatures the PETE efficiency is lower than the SQ limit due to the low thermal contribution. However, as the concentration increases, increased thermalization and absorption in the IR coupling element increase the cathode temperature and the thermally driven emission. Since the anode temperature is kept constant, the temperature difference between the PETE electrodes is increased, and the PETE efficiency rises above the SQ limit. As in the isothermal PETE, increasing the electrodes aspect ratio *S* reduces the radiative losses and further increases the efficiency. For example, at flux concentration of 1000, the efficiency reaches 45%, 49% and 52% for aspect ratios of 1, 5 and 20, respectively. The corresponding cathode temperatures are 1409 K, 1348 K and 1295 K, respectively.

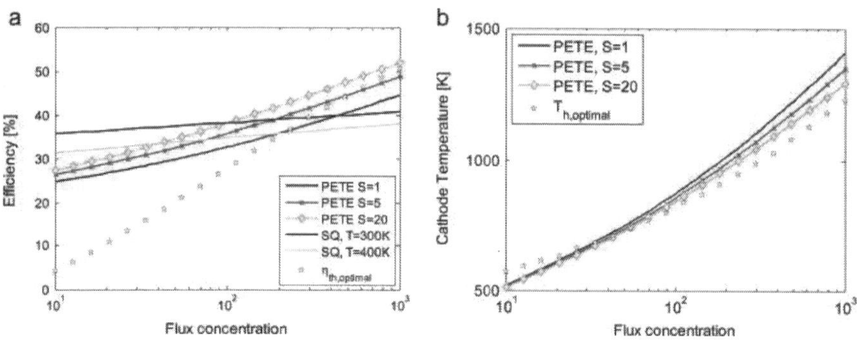


**Figure 7. (a)** Efficiency of non-isothermal PETE, ideal thermal, and SQ as a function of the flux concentration; (b) PETE cathode temperature according to the thermal balance and optimal solar thermal heat reservoir temperature. The cathode is doped $10^{19}$ $cm^{-3}$ p-type; the band gap is 1.4 eV and its electron affinity is 0.4 eV; the anode work function is 0.9 eV and its temperature is 500 K. The ambient temperature in the ideal solar thermal efficiency is 500 K.

The efficiency of an ideal solar thermal energy converter, as defined by Eq. (1), is also shown in Fig. 7(a). The receiver temperature was chosen to maximize the conversion efficiency, and the low temperature heat sink is at 500 K, identical to the anode temperature in the PETE case. For low flux concentrations, the ideal thermal converter operates at low temperatures, and produces lower efficiency than the PETE converter, which is driven mainly by direct photonic conversion. However, the thermal cycle efficiency increases rapidly with flux concentration and temperature, reaching 0.51 at flux concentration of 1000 and receiver temperature of 1230 K. The efficiency of a PETE converter under the same concentration and *S*=1 is lower than the ideal thermal converter, due to the high radiative recombination loss as discussed above. However, increasing the cathode aspect ratio reduces the recombination loss, and the efficiency of PETE with $S$=20 is higher than the ideal thermal cycle efficiency for the entire flux concentration range, reaching 0.52 at concentration of 1000. Therefore, the non-isothermal PETE converter can exceed, under some conditions, both the SQ limit and the ideal thermal converter limit.

It should be noted that at flux concentration of 1000, the thermal balance temperature of the cathode is in a range that may be impractical for real materials. However, the idealized model assumes that the cathode is completely isolated from its surroundings, while in any real system there will be some conduction heat transfer to the surroundings. Therefore, this value is the merely the upper bound for the cathode temperature. In any case, temperature limits of materials and thermal management are important considerations in addition to device efficiency.

Fig. 8 shows the efficiency of a two-stage non-isothermal PETE device with a secondary thermal converter as a function of the flux concentration and for different values of $S$. The PETE device parameters and cathode temperatures are as in Fig. 7. The low temperature heat sink for the secondary thermal cycle is 300 K. The secondary thermal cycle power and total efficiency were calculated according to Eqs. (23) and (28), where the currents and voltages are at the PETE maximum power point. Since the maximum power point voltage is higher than the difference in work functions [12], the heat flux density to the anode is the maximum power point current density multiplied by the anode work function. The maximum power point current density is therefore approximately linear with the flux concentration, and as a result the secondary thermal cycle efficiency is nearly constant with flux concentration, and it contributes about 11% of additional output. The total efficiency of the non-isothermal PETE with a secondary converter reaches 55%, 60% and 63% for cathode aspect ratios of 1, 5, and 20, respectively.

Fig. 8 shows also the efficiency of an ideal solar thermal converter with heat sink at 300 K. The ideal thermal efficiency increases with flux concentration due to the increase in receiver temperature, and it is higher than the two-stage PETE efficiency for flux concentrations above 140. This is different from the isothermal case, where the two-stage isothermal PETE efficiency was higher than the ideal thermal efficiency for all concentrations. This is due to the difference in heat transport to the secondary cycle: in the isothermal case, both the amount and the temperature provided to the secondary cycle are higher, and therefore its contribution can increase at higher concentrations. In the isothermal case, the heat flux to the secondary cycle includes heat generated by electron thermalization in both the cathode and anode thermalization, vs. anode only in the non-isothermal case. Furthermore, as seen in Fig. 6(b), the temperature that drives the secondary converter is higher compared to a fixed temperature of 500 K in the non-isothermal case.

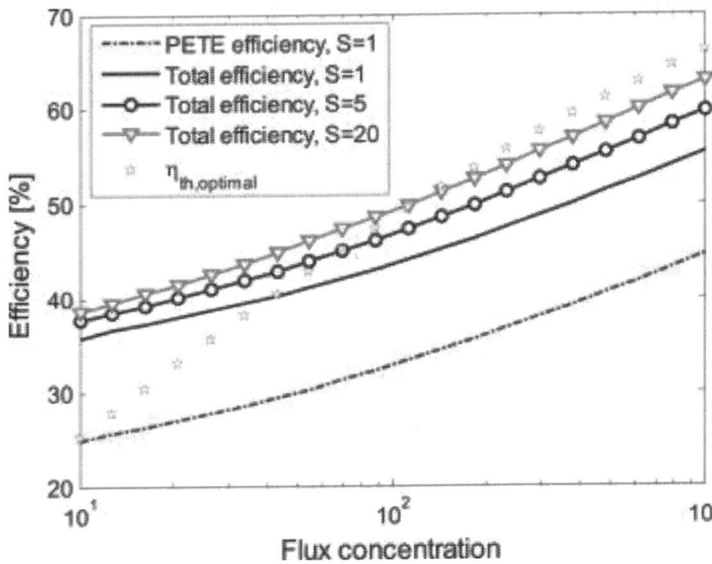


**Figure 8.** PETE efficiency, PETE with secondary thermal cycle efficiency and ideal solar thermal efficiency as a function of the flux concentration. The PETE device parameters and cathode temperatures are as in Fig. 7. The ambient temperature in the ideal solar thermal and secondary thermal cycle is 300 K.

## 3.4. Dual junction PETE, thermally coupled cathodes

When the two cathodes are optically and electrically separated, as shown in Fig. 2(a), the current in each sub-device depends on its operating voltage, which is independent from the voltage of the other sub-device, and on the shared temperature. The MPP of the full device must be found by optimization over two independent variables: $V_1$ and $V_2$, the operating voltages of the sub-device with higher bandgap and lower bandgap, respectively. It should be noted that the MPP of the full device, and the maximum power points of each sub-device, can occur at different combinations of voltages. The contour plots in Fig. 9(a) and (b) show an example of the variations of the power density output for each sub-device as a function of the voltages $V_1$ and $V_2$. The cathodes are $10^{19}$ $cm^{-3}$ p-type doped, the band gaps are 1.8 eV and 1.185 eV, and their electron affinity is 0.4 eV. The anodes work function and temperature are 0.9 eV and 500 K respectively, and the flux concentration is 1000. The difference between two adjacent iso-power density lines is 2.5 $W/cm^2$. The temperature that is reached as a result of the thermal balance for each combination of voltages is shown as a contour plot in Fig. 9(c). The difference between two adjacent iso-temperature lines is 100 K. In this example, the maximum output of the high band gap cathode is 28.55 $W/cm^2$ at voltages of $V_1$=1.81 V and $V_2$=0.06 V and temperature 1380 K. The maximum output of the low band gap cathode is 22.7 $W/cm^2$ at voltages of $V_1$=0.35 V and $V_2$=1.23 V and temperature 1320 K. The power density output of the entire device is shown in Fig. 9(d). The difference between two adjacent iso-power density lines is 2.5 $W/cm^2$. The maximum power output of the entire device is 46.1 $W/cm^2$at voltages of $V_1$=1.6 V and $V_2$=1.1 V and temperature 1153 K. Clearly, the MPP of the system is very different from the individual maximum power points of the two sub-devices. This behavior can be understood when noticing that the maximum output of each sub-device occurs when the other sub-device produces minimal electrical power, converting much of the incident radiation to thermal power, and increases the temperature and power output of the first sub-device. Therefore the MPP of the entire device is an intermediate point that balances the output of the two sub-devices.

Fig. 10 presents contour plots of the performance of the dual band PETE device with thermally coupled cathodes as a function of the bandgaps of the two cathodes: the efficiency of each sub-device separately (a,b), the cathode temperature (c), and the overall device efficiency (d). All parameters are as in Fig. 9. The difference between adjacent iso-efficiency lines is 2.5% (Fig. 10 a,b,d) and between two adjacent iso-temperature (Fig. 10c) line is 100 K. The maximum power point for each two bandgaps was taken to be the maximum power point of the complete device. For high $Eg_{,1}$, the low band gap cathode absorbs larger parts of the solar spectrum, and its performance approaches that of an individual PETE cathode with an IR coupling element. At the maximum efficiency of the low band gap sub-device reaches 35.7% with , and increasing $E_{g1}$ even further will bring the low band gap sub-device performance to the values of the single cathode PETE, as discussed above. The high band gap cathode reaches a maximum efficiency of 44.9% for $Eg_{,1}$=1.33 eV and $Eg_{,2}$=1.3 eV and temperature of

1357 K. In this case, the low band gap cathode absorbs mostly through the IR coupling element, and operates as a thermionic converter without much photo-enhancement. The maximum efficiency of the complete device is 51.2% for band gaps of 1.807 eV and 1.18 eV, and the cathodes temperature is 1156 K. The contribution of the two sub-devices is well balanced, with 22.5% and 28.7% for the low band gap cathode and the high band gap cathode, respectively.

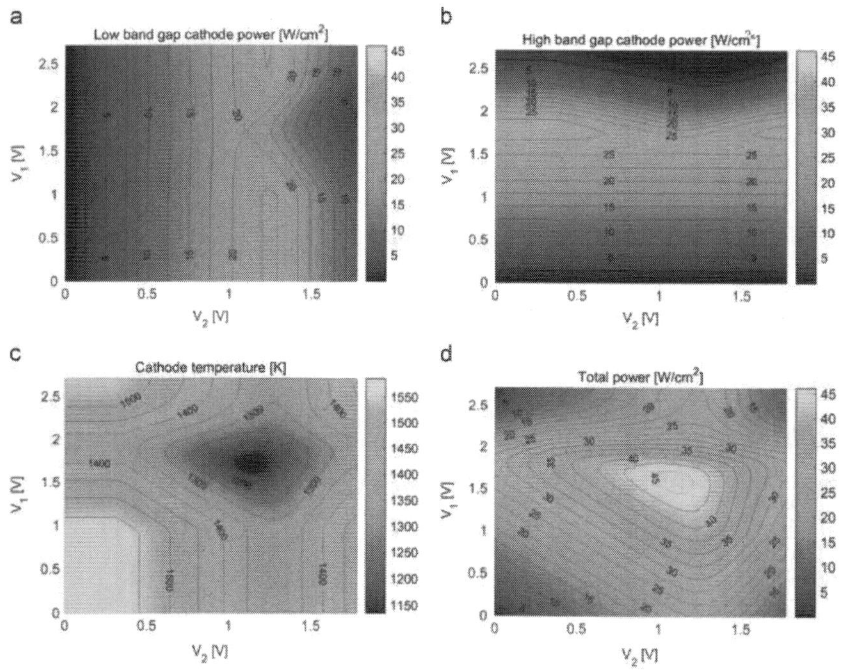


**Figure 9. (a)** Low band gap sub device power density, (b) high band gap sub device power density, (c) cathode temperature, and (d) total power density, as a function of the sub-devices operating voltages. The cathodes are thermally coupled, $10^{19}$ $cm^{-3}$ p-type doped, their electron affinity is 0.4 eV, the anodes work function and temperature are 0.9 eV and 500 K respectively, the band gaps are 1.8 eV and 1.185 eV, and the flux concentration is 1000. The difference between two adjacent iso-power lines is 2.5 $W/cm^2$ (a,b,d) and between two adjacent iso-temperature line is 100 K.

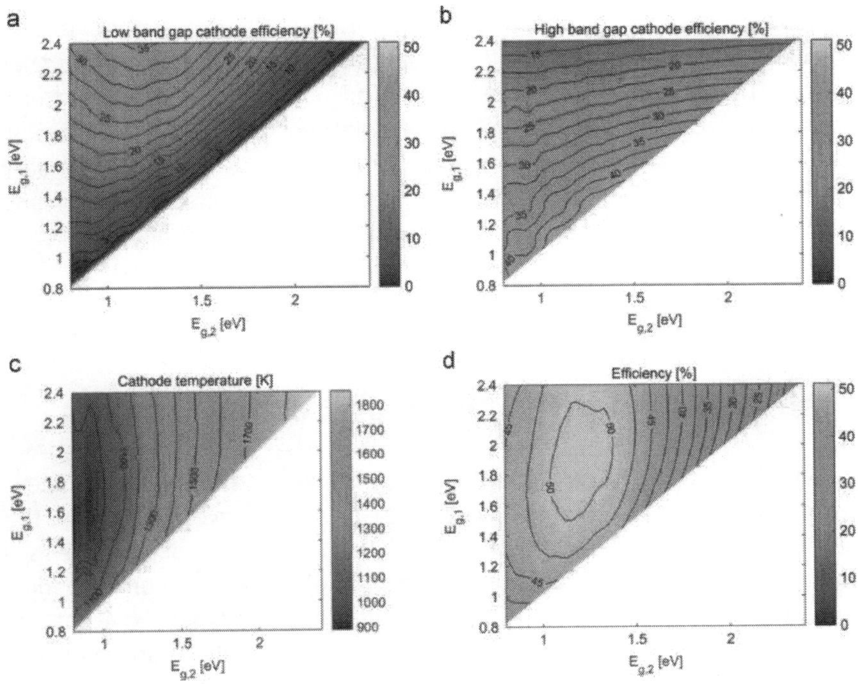


**Figure 10. (a)** Low band gap sub device efficiency, (b) high band gap sub device efficiency, (c) cathodes temperature, and (d) full device efficiency, as a function of the cathodes band gaps. The cathodes are thermally coupled, and all parameters are as in Fig. 9. The difference between adjacent iso-efficiency lines is 2.5% (a,b,d) and between two adjacent iso-temperature (c) line is 100 K.

Increasing the electrodes aspect ratio *S* can increase the efficiency of dual junction PETE devices, similar to the case of the single bandgap PETE devices. For example, for a dual bandgap converter under the conditions defined in Fig. 9 with an aspect ratio of *S*=20, the maximum efficiency increases to 56.3%.

In the dual junction configurations described above, heat can extracted from both anodes and converted in a separate thermal cycle. For example, for a dual band gap PETE device with the same parameters as in Fig. 9, feeding all anode thermalization heat into an ideal heat engine, an additional 13.8% of the incident solar power can converted into electricity, leading to a total system efficiency of 65.1%. Adding a secondary thermal cycle to a thermally coupled dual band gap system with an aspect ratio of

20 yields a total efficiency of 70.4%. These results correspond to the specific anode temperature used here, and it should be possible to further increase the overall efficiency by seeking an optimal anode temperature. An optimum is expected due to the following opposite effects: increasing the anode temperature will increase the efficiency of the secondary thermal cycle, but will also increase the anodes reverse emission and reduce the PETE efficiency, and vice versa. This additional optimization is left for future work.

## 3.5. Dual junction PETE, thermally isolated cathodes

The efficiency of a dual junction PETE where a separate thermal balance is applied to each cathode was calculated for a wide range of band gaps. Fig. 11(a,b) shows the efficiency for each of the sub-devices separately as a function of the cathodes band gaps. Fig. 11(c,d) shows the corresponding temperatures of the two cathodes. The difference between adjacent iso-efficiency lines is 2.5% (Fig. 11a,e) and between two adjacent iso-temperature (Fig. 11c) line is 100 K. Since the high band gap cathode does not have any interaction with sub-band gap radiation, its temperature and efficiency are independent of the lower band gap of the second cathode. The high bandgap cathode operates as a single PETE device without an IR coupling element, and yields maximum efficiency of 40.5% at temperature of 1185 K and band gap of 0.95 eV. The optimal band gap and efficiency are lower than those shown in Section 3.3 above, because the lack of an IR coupling element moves the optimum towards increased absorption of low energy photons, in order to increase the cathode temperature.

Fig. 11(e) shows a contour plot of the total efficiency of the combined converter device as a function of the two cathodes band gaps. The electrodes parameters are as in Fig. 9 and the flux concentration is 1000. The complete device efficiency reaches 50.3% where the high band gap cathode contributes 25.4% at $Eg_{,1}$=1.85 eV and temperature of 1547 K and the lower band gap cathode contributes 24.9% at $Eg_{,2}$=1.15 eV and temperature of 1149 K (Fig. 11e). Adding a secondary thermal cycle with a Carnot efficiency at as described in Section 3.5 to a system with $Eg_{,1}$=1.85 eV and $Eg_{,2}$=1.15 eV brings the efficiency to 66.8%.

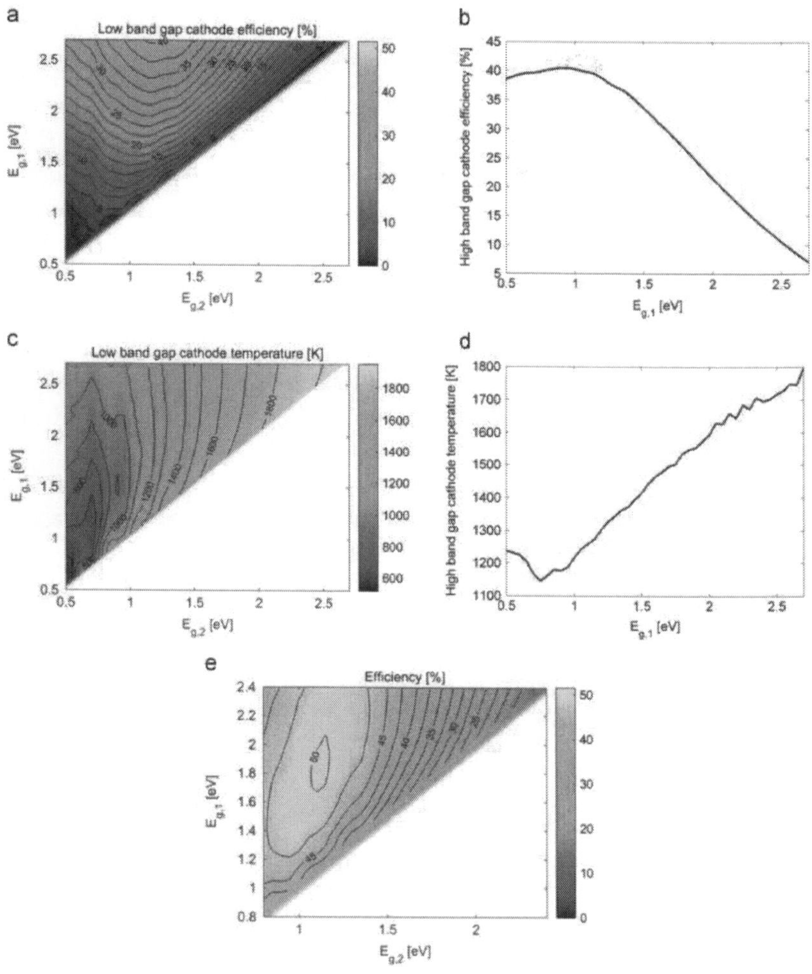


**Figure 11.** Performance as a function of the cathodes band gaps (a) low band gap sub-device efficiency, (b) high band gap sub device efficiency, (c) low band gap cathode temperature, (d) high band gap cathode temperature, and (e) total device efficiency. The cathodes are thermally isolated and all parameters are as in Fig. 9. The difference between adjacent iso-efficiency lines is 2.5% (a,e) and between two adjacent iso-temperature (c) line is 100 K.

As in the previous examples, increasing the electrodes aspect ratio can further increase the conversion efficiency. For example, a system with $E_{g,1}$=1.85 eV and $E_{g,2}$=1.15 eV and an aspect ratio of 20 yields an efficiency of 54.3% and 69% with the addition of a secondary thermal cycle.

# 4. SUMMARY AND DISCUSSION

A summary of the theoretical efficiencies of the different converter configurations for concentration of 1000 is given in Table 1. The corresponding SQ limit (at room temperature) is 41% for a single junction cell, and 54.4% for a dual junction cell. The non-isothermal PETE converter efficiency exceeds the single-junction SQ limit, with a significantly higher advantage as the geometric aspect ratio increases, due to the reduction in radiative recombination loss as described above. Adding the secondary thermal stage increases the advantage of PETE conversion even further, reaching 63% for high aspect ratio, compared to only 41% that can be achieved by an ideal single-junction PV cell under the same concentration. Note that the use of waste heat from the PV cell to augment electrical output is not possible when the cell is at ambient temperature, and not practical at the high temperatures that might make a significant impact [9].

**Table 1.** Efficiency limits of the different PETE converter configurations under concentration of 1000.

| Converter | $S=1$ | | $S=20$ | |
|---|---|---|---|---|
| | PETE (%) | +Secondary (%) | PETE (%) | +Secondary (%) |
| Isothermal | 24.7 | 69.8 | 24.7 | 69.8 |
| Non-isothermal | 44.7 | 55.5 | 52.1 | 63.0 |
| Dual, coupled | 51.2 | 65.1 | 56.3 | 70.4 |
| Dual, isolated | 50.3 | 66.8 | 54.3 | 69.0 |

The efficiency of the isothermal PETE converter is bounded by the SQ limit that corresponds to the operating temperature, and therefore it is much lower than the SQ limit at room temperature. However, when adding the contribution of a secondary thermal converter, the combined device can significantly exceed the SQ limit, reaching 69.8% under the most favorable conditions. The two-stage efficiency of the isothermal device is also higher than the non-isothermal case. This is due to the fact that in the isothermal configuration more heat is available to the secondary converter (collected from thermalization energy of both anode and cathode), and the driving temperature of the secondary converter is higher. Concurrently, heat loss to the environment by blackbody emission is lower than the non-isothermal cases due to the lower cathode temperature.

The efficiency of the dual bandgap PETE devices is higher than single bandgap PETE. However, the improvement due to spectral splitting is modest, in contrast to PV cells where the improvement in dual junction vs. single junction is more dramatic. This can be explained by recalling that spectral splitting is a method to reduce the thermalization loss. However, PETE converters utilize the thermalization energy to increase the cathode temperature and increase electron emission, and therefore thermalization energy is at least partially recovered. Reducing thermalization by spectral splitting will have then a smaller impact on PETE converters efficiency, in contrast to PV devices. Furthermore, since the input solar radiation is divided between two cathodes, the operating temperature of each cathode is lower than the operating temperature of the single band gap PETE cathode, reducing the contribution of thermal conversion. As a result, the dual bandgap PETE converter efficiency is somewhat lower than (for $S$=1) or close to ($S$=20) the SQ limit for a dual junction PV cell. Similar to the single bandgap case, adding the secondary thermal converter leads to a significant advantage of the PETE device, while this kind of addition is not available for PV cells.

The ideal thermal conversion limits corresponding to concentration of 1000 are 51% and 66%, for heat sink temperatures of 500 K and 300 K, respectively. The first case is relevant for comparison to a PETE-only converter with the same anode temperature, while the second is appropriate for comparison against the two-stage PETE with a secondary converter having the environment as its heat sink. Table 1 shows that the single bandgap non-isothermal PETE converter does not exceed the thermal limit, but the isothermal and dual bandgap versions with high aspect ratio can produce efficiency higher than this limit. It is interesting to note that the isothermal PETE converter, while offering lower efficiency in a stand-alone configuration, can exceed the thermal limit when configured with a secondary thermal stage. It is remarkable that this high overall efficiency corresponds to a device temperature of 920 K that is much lower than the optimal cathode temperatures for the other configurations, making this case much more accessible for practical implementation.

Reck and Hansen [10] have shown that the ideal PETE efficiency including a secondary thermal stage can reach 58% at flux concentration of 1000, band gap of 0.56 eV, and anode work function of 0.41 eV. This solution requires anode temperature of 328 K, leading to a very low contribution of the thermal stage. It also corresponds to no quasi-Fermi level splitting, i.e., the device operates as a purely thermionic converter rather than a PETE device. This solution cannot be directly compared with the current results, due to two significant differences in the analysis. First, anode work function of 0.41 eV cannot be achieved with any known material, and we have used the lowest value reported in the literature of 0.9 eV. Second, Reck and Hansen assumed that only above bandgap radiation is absorbed and emitted, while we have assumed that there is an IR coupling element that allows radiation absorption and emission over the entire spectrum. A PETE device with an IR coupling element, operating under the same conditions but with a band gap of 1.4 eV will produce the same efficiency as calculated by [10] with non-zero quasi-Fermi level splitting. This indicates that the introduction of an IR coupling element has a very significant effect on the device operation.

We have defined here an IR coupling element that absorbs and emits over the entire spectrum. However, this is not necessarily the optimal definition. If we assume a selective IR absorber with emissivity one in a specific part of the solar spectrum (where incident sunlight is available), and emissivity zero in the rest of the spectrum (where sunlight is negligible but thermal emission is significant), then even higher performance might be achieved. This is similar to the selective coatings used very effectively in solar thermal applications. Optimization of such a selective IR element is left for future work.

An additional efficiency increase can be achieved by optimizing the anode temperature (in non-isothermal configurations) and anode work function in conjunction with the secondary thermal cycle. As discussed above, the heat flux to the secondary cycle is determined by the anode work function. Hence, a lower work function increases the PETE efficiency but reduces the heat flux to the secondary cycle, and vice versa. Similarly, higher anode temperature can reduce the PETE efficiency but will increase the secondary thermal cycle efficiency, and vice versa. Furthermore, since the power output to the anode depends also on the device current density, the PETE-only maximum power point is not necessarily the maximum power point of the entire two-stage device. Optimization of the anode and secondary cycle parameters are left for future work.

In this work we have assumed that the cathode contact is ideally selective, i.e., it is completely transparent to holes and does not transmit nor removes electrons from the cathode. Recent work [16] has shown that under some operating conditions electrons are injected from the contact into the anode, and may increase the net current density beyond the value that will be achieved with an ideal selective contact. This effect is not considered here, and future work should consider how this should be included in the analysis, and how much it may contribute to the limit on conversion efficiency.

Some of the PETE converter parameters were set as constant in the current work, but variations in these parameters may further increase the conversion efficiency limit. The cathode doping was taken to be $10^{19}$ $cm^{-3}$ as this value is close to the maximum p-type doping that is possible with standard fabrication processes. Reducing this value can increase the emission current density, but will also reduce the MPP voltage which is close to the difference in the electrodes work functions. Therefore the overall impact on the conversion efficiency should be examined. The electron affinity was set as 0.4 eV, and variation of this property probably will not have a significant effect, as long as it is low enough [12]. The anode work function chosen in this work is 0.9 eV, corresponding to the lowest work function realized experimentally [15]. If lower work function anode could be realized, the conversion efficiency of PETE devices could be further increased as shown in [7].

Practical implementation of the suggested devices may be difficult due to availability of materials with the specific band gap requirements which are capable of withstanding such high temperatures. In multi-junction solar cells band gap tuning is mostly done by modifying the mole fraction of III–V semiconductor alloys. However, there is very little data on their electronic properties at elevated temperatures. On the other hand, some materials are suggested for high temperature electronics such as diamond or silicon carbide, but their bandgap is too high to allow efficient solar energy conversion. The high temperature challenge applies not only to the active cathode material, but also to additional elements such as electrical contacts, window layer, etc., that have to coexist and operate reliably at high temperature. Development of material systems with appropriate band gaps and other required electronic properties, and with robustness and mutual compatibility at elevated temperatures, is a key step needed towards realization of high efficiency PETE solar energy converters.

# 5. CONCLUSIONS

The upper limit on conversion efficiency of PETE devices was analyzed for several possible device configurations. The main conclusions can be summarized as follows:

- The efficiency of PETE converters can exceed under some conditions both the SQ limit and the ideal thermal limit. This represents the hybrid nature of PETE conversion that involves a synergy of both photonic and thermal processes.
- Spectral splitting can increase the PETE efficiency, but to lesser extent compared to spectral splitting in tandem PV cells. This is since thermalization energy is already utilized in PETE even in a single bandgap configuration.
- The contribution of a secondary thermal cycle is substantial for all PETE converter configurations. The highest conversion efficiency reported here, 70.4%, is largely due to the significant contribution of the thermal stage.
- Several device parameters were not fully explored in this work, and some of them may hold potential to increase the efficiency limits even further. Additional work is then needed to complete the mapping of the performance envelope of PETE converter devices.
- The optimal operation temperatures for all non-isothermal configurations (single and dual bandgap) are well above the range where semiconductors are known to operate in practical devices. Exploring the temperature limits of real materials is then a major challenge in developing real PETE devices.

We have shown that PETE conversion can compete well against the mainstream paths of photovoltaic and solar thermal electricity generation, at least on the basis of theoretical efficiency limit. How close real PETE devices may approach these limits remains to be seen in the future, as the PETE technology is developed and demonstrated.

## ACKNOWLEDGMENTS

This work was supported by the European Union's Seventh Framework Programme for Research, Technological Development and Demonstration under Grant Agreement no.308975 (project PROMETHEUS, http://www.prometheus-energy.eu/). G. Segev acknowledges the support of Eshkol fellowship from the Israel Ministry of Science and Technology under Grant Agreement no. 3-8638.

## REFERENCES

1. W. Shockley, H.J. Queisser, Detailed balance limit of efficiency of p–n junction solar cells, J. Appl. Phys. 32 (3) (1961) 510–519.

2. Martí, G. Araújo, Limiting efficiencies for photovoltaic energy conversion in multigap systems, Sol. Energy Mater. Sol. Cells 43 (1996) 203–222.

3. M.A. Green, Third Generation Photovoltaics: Advanced Solar Energy Conversion, Springer, Berlin, Heidelberg, 2006.

4. E.A. Fletcher, R.L. Moen, Hydrogen and oxygen from water, Science 197 (1977) 1050.

5. G.N. Hatsopoulos, E.P. Gyftopoulos, Thermionic Energy Conversion, MIT Press, Cambridge, Massachusetts, 1973.

6. J.W. Schwede, I. Bargatin, D.C. Riley, B.E. Hardin, S.J. Rosenthal, Y. Sun, F. Schmitt, P. Pianetta, R.T. Howe, Z.X. Shen, N.A. Melosh, Photon-enhanced thermionic emission for solar concentrator systems, Nat. Mater. 9 (9) (2010) 762–767.

7. G. Segev, A. Kribus, Y. Rosenwaks, High performance isothermal photo-thermionic solar converters, Sol. Energy Mater. Sol. Cells 113 (2013) 114–123.

8. Y. Yang, W. Yang, W. Tang, C. Sun, High-temperature solar cell for concentrated solar-power hybrid systems, Appl. Phys. Lett. 103 (8) (2013) 083902.

9. Kribus, G. Mittelman, Potential of polygeneration with solar thermal and photovoltaic systems, J. Sol. Energy Eng.: ASME 130 (1) (2008).

10. K. Reck, O. Hansen, Thermodynamics of photon-enhanced thermionic emission solar cells, Appl. Phys. Lett. 104 (2) (2014) 023902.

11. S. Meir, C. Stephanos, T.H. Geballe, J. Mannhart, Highly-efficient thermoelectronic conversion of solar energy and heat into electric power, J. Renew. Sustain. Energy 5 (4) (2013) 043127.

12. G. Segev, Y. Rosenwaks, A. Kribus, Efficiency of photon enhanced thermionic emission solar converters, Sol. Energy Mater. Sol. Cells 107 (2012) 125–130.

13. Varpula, M. Prunnila, Diffusion–emission theory of photon enhanced thermionic emission solar energy harvesters, J. Appl. Phys. 112 (4) (2012) 044506.

14. P. Wurfel, Physics of Solar Cells from Principles to New Concepts, Wiley, Weinheim, 2005.

15. F.A.M. Koeck, R.J. Nemanich, A. Lazea, K. Haenen, Thermionic electron emission from low work-function phosphorus doped diamond films, Diam. Relat. Mater. 18 (5–8) (2009) 789–791.

16. G. Segev, Y. Rosenwaks, A. Kribus, Loss mechanisms and back surface field effect in photon enhanced thermionic emission converters, J. Appl. Phys. 114 (2013) 044505.

# Chapter 4

# Towards 12% stabilised efficiency in single junction polymorphous silicon solar cells: experimental developments and model predictions

**Sergey Abolmasov[1,2a], Pere Roca i Cabarrocas[2] and Parsathi Chatterjee[2]**

[1] R&D Center of Thin-Film Technologies in Energetics, Ioffe Institute 28 Polytekhnicheskaya, 194064 Saint Petersburg, Russia

[2] LPICM, CNRS, Ecole Polytechnique, Université Paris-Saclay, 91128 Palaiseau, France

## ABSTRACT

We have combined recent experimental developments in our laboratory with modelling to devise ways of maximising the stabilised efficiency of hydrogenated amorphous silicon (a-Si:H) PIN solar cells. The cells were fabricated using the conventional plasma enhanced chemical vapour deposition (PECVD) technique at various temperatures, pressures and gas flow ratios. A detailed electrical-optical simulator was used to examine the effect of using wide band gap P-and N-doped $\mu$c-$SiO_x$:H layers, as well as a $MgF_2$ anti-reflection coating (ARC) on cell performance. We find that with the best quality a-Si:H so far produced in our laboratory and optimised deposition parameters for the corresponding solar cell, we could not attain a 10% stabilised efficiency due to the high stabilised defect density of a-

Si:H, although this landmark has been achieved in some laboratories. On the other hand, a close cousin of a-Si:H, hydrogenated polymorphous silicon (pm-Si:H), a nano-structured silicon thin film produced by PECVD under conditions close to powder formation, has been developed in our laboratory. This material has been shown to have a lower initial and stabilised defect density as well as higher hole mobility than a-Si:H. Modelling indicates that it is possible to attain stabilised efficiencies of 12% when pm-Si:H is incorporated in a solar cell, deposited in a NIP configuration to reduce the P/I interface defects and combined with P- and N-doped $\mu$c-SiO$_x$:H layers and a $MgF_2$ ARC.

## 1 INTRODUCTION

Large-scale application of amorphous silicon based solar cells in the photovoltaic power generation industry requires improvement of their efficiency. Hydrogenated amorphous silicon (a-Si:H) based solar cells, in spite of their advantage of large area deposition, lag behind their crystalline counterparts with respect to conversion efficiency. Moreover they suffer from light-induced degradation, which can reduce cell efficiency by ~25%. Therefore much efforts have been made to develop growth conditions leading to a-Si:H based materials having improved transport properties or stability e.g. [1, 2]. Also tandem or triple-junction solar cells, with thin, varying band gap materials in different sub-cells have been tried out and achieved an initial record efficiency of 16.3% [3]. This design improves light absorption over the entire visible spectrum and reduces light-induced degradation as it is possible here to use thinner sub-cells. However the design is complex and it would be ideal technologically to attain a high stabilised efficiency in single junction solar cells. Moreover, this is a requisite for high stabilized efficiency in tandem and triple junction solar cells.

To this end, we have been experimenting with an a-Si:H-like material: hydrogenated polymorphous silicon (pm-Si:H), which is produced by the standard RF glow discharge decomposition of silane highly diluted in hydrogen, under conditions close to powder formation [4, 5, 6, 7, 8]. Pm-Si:H is a nano-structured silicon material with medium range order which shows up in the superior hole transport properties as deduced from diffusion induced time resolved microwave conductivity [8] and time-of-flight measurements [9]. Moreover it exhibits lower initial [5] and stabilised defect densities than standard a-Si:H [1, 9,10], with the $\eta\mu\tau$ product of electrons 200–700 times higher than that of standard a-Si:H for samples in the as-deposited state [1], and attaining values after light-soaking comparable to those of standard a-Si:H films before degradation [1, 4, 5, 6,7]. In reference [1], a comparative study has been made of the kinetics of defect creation in standard a-Si:H and pm-Si:H films. It is found that the value of the absorption coefficient $\alpha$(1.1 eV), obtained at 'saturation' is lower for the pm-Si:H than standard a-Si:H films, indicating a lower stabilized density of states (DOS) for the former case.

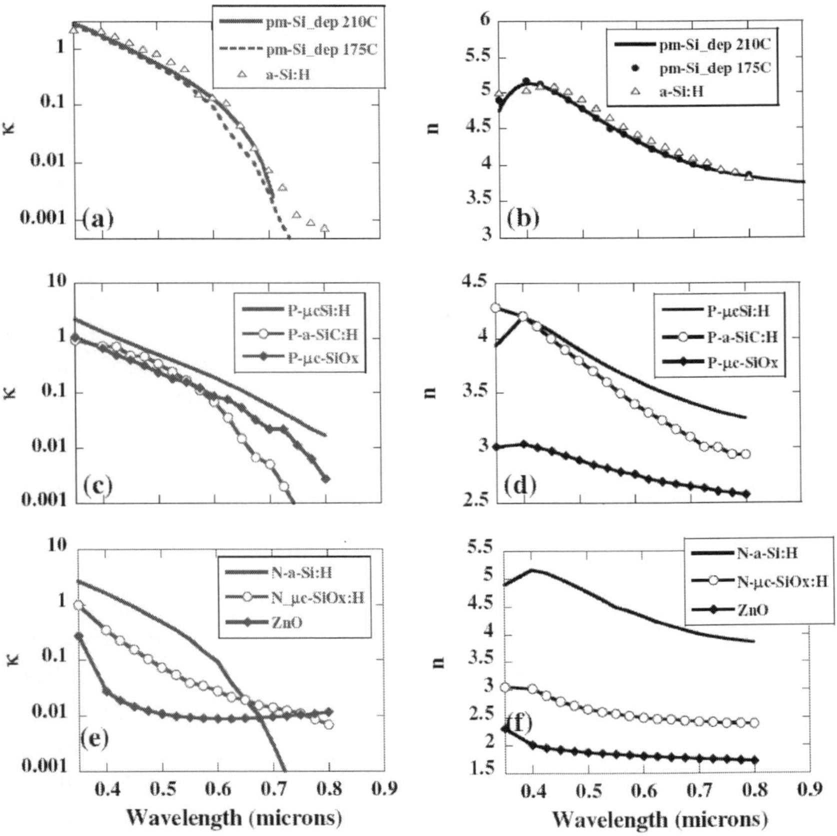


**Figure 1**. Comparison of the complex refractive indices ($\kappa$, $n$) (a) and (b) of the intrinsic layers used in this study: a-Si:H, pm-Si:H deposited at 175 °C and 210 °C; (c) and (d) of the various P-layers studied: P-$\mu$c-Si:H, P-a-SiC:H and P-$\mu$c-$SiO_x$:H; and (e) and (f) of the different N-layers: N-a-Si:H, N-$\mu$c-$SiO_x$:H and the TCO: ZnO:Al.

Unfortunately the material has slightly lower absorption coefficients than a-Si:H, therefore lower coefficients of extinction ($\kappa$ – Fig. 1a) over the visible spectrum [11]. The wavelength dependence of the real part of the complex refractive indices (CRINDs) of pm-Si:H and a-Si:H are given in Figure 1b. Also in the same article [11], using detailed modelling to simulate the measured characteristics of solar cells fabricated with this material as the intrinsic layer, we have shown that pm-Si:H develops a very defective P/I interface, probably due to bombardment by charged silicon nanocrystals contributing to the deposition and due to the high flux of hydrogen required to develop the material. Moreover other experiments performed at our laboratory [12] have shown that light-soaking of pm-Si:H leads to the formation of hydrogen bubbles that introduce macroscopic defects. In this article we show by modelling how the above drawbacks of a pm-Si:H solar cell can be overcome to produce solar cells with a high stabilised efficiency.

We have also tried out different window structures using modelling, to see which combination leads to the highest efficiency. Two types of front window designs, each with two P-layers were examined: (a) P-$\mu$c-Si:H/P-a-SiC:H or (b) P-$\mu$c-Si:H/P-$\mu$c-$SiO_x$. In a previous paper [13] modelling the effects of P-$\mu$c-Si:H, P-a-SiC:H window layers and various combinations thereof, we have shown that a P-$\mu$c-Si:H/P-a-SiC:H stack is an efficient design. Indeed, the $\mu$c-Si:H P-layer has a lower surface band bending (sbb) and lower activation energy than P-a-SiC:H. These properties help to improve the built-in potential ($V_{bi}$), open-circuit voltage ($V_{oc}$) and fill factor (FF) of solar cells employing such a double window layer structure relative to a standard cell having a single P-a-SiC:H window layer [13]. These modelling results have received confirmation from the experimental work of Ma et al. [14]. The problem with a single P-$\mu$c-Si:H window is the large valence band discontinuity, since it has been shown by the IPE measurements of Xu et al. [15] that the P-$\mu$c-Si:H/a-Si:H band discontinuity is almost totally on the valence band side. This fact results in a large drop in potential at the P/I interface, leading to a collapse of the electric field inside the volume of the device and reducing the $V_{oc}$ and the FF of a device employing such a single P-$\mu$c-Si:H window. Also considerable back diffusion of electrons takes place, reducing the blue-green part of the external quantum efficiency (EQE) and short-circuit current density ($J_{sc}$). In addition the P-$\mu$c-Si:H layer developed in our laboratory has a higher coefficient of extinction (hence higher absorption coefficient) compared to P-a-SiC:H or P-$\mu$c-$SiO_x$:H over the visible solar spectrum, as seen from Figure 1c. A thin P-a-SiC:H buffer sandwiched between the P-$\mu$c-Si:H and the a-Si:H layers solves all these problems, since the band discontinuity between a-Si:H and P-a-SiC:H is mainly on the conduction band side [16], preventing back diffusion of electrons and bringing up the short wavelength QE and $J_{sc}$. Also, the detrimental effect of the large valence band discontinuity is reduced by the presence of a thin P-a-SiC:H layer [13], leading to high values of $V_{oc}$ and FF. Moreover the presence of P-a-SiC:H, allows to use a thinner P-$\mu$c-Si:H layer, so that the absorption of the shorter wavelengths in the P-layers is reduced. P-doped $\mu$c-$SiO_x$:H is also a promising P-layer, since it is more transparent (having a wider band gap) than P-a-SiC:H and its doping density can be adjusted to yield high conductivity. We have in this study therefore also tried out by modelling a P-$\mu$c-Si:H/P-$\mu$c-$SiO_x$:H double window structure, primarily because we know the TCO/P sbb for the case of P-$\mu$c-Si:H [14], but it is unknown in the case of P-$\mu$c-$SiO_x$:H. Nevertheless we have also experimented with a single P-$\mu$c-$SiO_x$:H window layer, assuming a range of values for the ZnO:Al/P-$\mu$c-$SiO_x$:H sbb. In all cases an ultra-thin intrinsic buffer layer follows the window P layers, both to prevent boron tailing into the intrinsic layer and to obtain a less defective P/I interface.

Different back contact designs were also investigated via modelling. This was either N-a-Si:H/ZnO:Al/Ag or simply N-$\mu$c-$SiO_x$:H/Ag with the ZnO:Al left out. The logic behind this latter design is that the band gap and doping density of N-$\mu$c-$SiO_x$:H can be adjusted in such a way that it can be used as a wide band gap TCO, while retaining a good enough conductivity to act as the N-layer of the cell [17]. Thus the deposition of an extra TCO layer can be avoided. However this is only possible in cases where the P-layer is deposited first on ZnO:Al coated glass. When a cell is deposited in a NIP configuration, a thin ZnO:Al layer between the silver and N-$\mu$c-$SiO_x$:H is mandatory to prevent silver diffusion into the latter. The CRIND's of N-$\mu$c-$SiO_x$:H are compared to those of standard N-a-Si:H and the TCO ZnO:Al in Figures 1e and 1f. Finally we have used modelling to study how a thin anti-reflection coating (ARC), deposited on the surface through which light enters the solar cell, can affect$J_{sc}$ (and hence the conversion efficiency $\eta$).

In brief, the aim of this article is to use the measured values of the initial and stabilised state properties of the best a-Si:H and pm-Si:H materials developed in our laboratory so far, together with realistic P/I interface defect density and other properties extracted by modelling the corresponding solar cells in an earlier article [11], combined with different emitter and N-layer materials and designs, to predict using modelling the highest efficiency attainable in these structures. We find, using the material and deposition parameters of our a-Si:H solar cells, that it is difficult to attain 10% stabilised efficiency with this material, even though other laboratories e.g. [18] have actually developed a-Si:H single junction cells with stabilised efficiencies in excess of 10%. This is likely due to the superior nature and/or light absorption of the a-Si:H material developed, improved P/I interface properties in their deposited solar cells, etc. On the other hand we have been able to develop pm-Si:H [1, 4, 5, 6, 7, 8], a close cousin of a-Si:H, with lower initial and stabilised defect densities, higher hole mobility, but lower light absorption. Using modelling we show in this article that such a material when deposited in a solar cell in the NIP configuration, and in conjunction with suitable P- and N-layers and an ARC is capable of attaining a stabilised efficiency of around 12%, although our best experimental results so far show a lower stabilised efficiency [19].

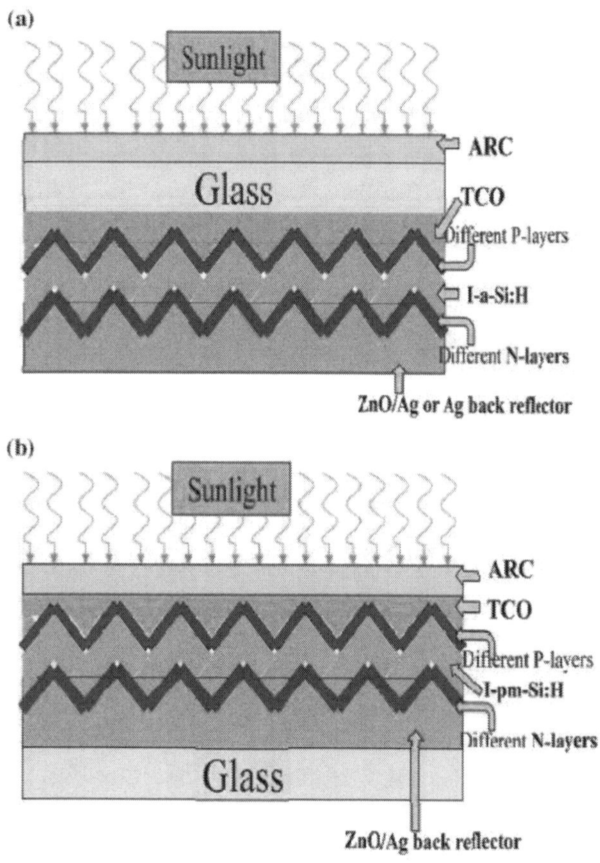


**Figure 2.** Schematic representations of (a) an a-Si:H solar cell (deposited in the PIN configuration on TCO coated glass) and (b) a pm-Si:H solar cell deposited in the NIP configuration on TCO/Ag coated glass.

Schematic diagrams of the a-Si:H solar cell deposited in a PIN configuration on TCO-coated glass, and of the pm-Si:H cell deposited in a NIP configuration on glass/Ag/ZnO:Al are shown in Figures 2a and 2b, respectively. These are the structures that have been modelled. It would have been better if the ARC were assumed to be textured also; however as will be pointed out in the section on the simulation model ASDMP, this latter at present is capable of handling only two textured interfaces. These have been taken at the TCO/P and N/back reflector (ZnO:Al/Ag or only Ag) interface to maximise light absorption in the intrinsic layer, except when $\mu$c-SiO$_x$:H is used as the N-layer. The refractive index of this N-layer is close to that of ZnO:Al, so that most of the light in reality suffers specular and diffused reflection at the I/N-$\mu$c-SiO$_x$:H interface. Hence it is this latter interface that is taken as the second textured interface in the modelling calculations. The layers between the top TCO and bottom reflector are thin and we assume that they almost exactly take up the structure of the preceding layer (conformal deposition). Therefore we have assumed that only specular reflection, obeying Fresnel's law, takes place at these interfaces.

## 2 EXPERIMENTS

The PIN devices were deposited at various temperatures ranging from 175 °C to 240 °C by the RF (13.56 MHz) glow discharge plasma enhanced chemical vapor deposition (PECVD) method in a multiplasma-monochamber reactor [20]. Standard a-Si:H films were produced by the dissociation of pure silane at 40 mTorr under an RF power of 10 mW/cm$^2$. On the other hand, pm-Si:H films were obtained by the dissociation of a 10% silane in hydrogen mixture at 2–4 Torr under an RF power of 90 mW/cm$^2$. A deposition rate of 9 Å/s could be achieved for the pm-Si:H films [21]. We experimented with two different deposition temperatures (175 °C and 210 °C), but were unable to obtain any significant increase in the absorption coefficients in the case of a-Si:H. However for pm-Si:H, a substrate temperature of 210 °C allowed to obtain films with a higher absorption over the visible wavelengths than for pm-Si:H deposited at 175 °C. The CRINDs of a-Si:H, and pm-Si:H deposited at two different temperatures are compared in Figures 1a and 1b.

**Table 1** The input parameters used to model different cases. The first 5 nm of the I-pm-Si:H layer at the N/I interface has a DB DOS of 5 × $10^{17}$ $cm^{-3}$, followed by a 30 nm intermediate layer with a DOS = 2 × $10^{16}$ $cm^{-3}$. The first 20 nm of the I-a-Si:H layer adjacent to the P/I interface is assumed to have a DOS of 2 × $10^{16}$ $cm^{-3}$. The values have been extracted from modelling previous experimental results [11]. The parameters of N-a-Si:H are not shown as these are standard.

| Parameter | P-$\mu c$-Si:H | P-a-SiC:H | P-$\mu c$-SiO$_x$ | N-$\mu c$SiO$_x$ | I-pm-Si:H | I-a-Si:H |
|---|---|---|---|---|---|---|
| Layer thickness (nm) | 6 | 4 | 10 | 10 | 250 | 200–300 |
| Electron affinity (eV) | 4.00 | 3.89 | 3.67 | 4.00 | 3.95 | 4.00 |
| Mobility gap (eV) | 1.6 | 2.0 | 2.5 | 2.4 | 1.96 | 1.84 |
| Activation energy (eV) | 0.12 | 0.46 | 0.47 | 0.09 | 0.92 | 0.86 |
| Eff. DOS in bands ($cm^{-3}$) | $2 \times 10^{20}$ | $2 \times 10^{20}$ | $2 \times 10^{20}$ | $2 \times 10^{20}$ | $2 \times 10^{20}$ | $2 \times 10^{20}$ |
| Ch Energy (VB tail) (eV) | 0.035 | 0.120 | 0.120 | 0.025 | 0.050 | 0.050 |
| Ch Energy (CB tail) (eV) | 0.025 | 0.07 | 0.070 | 0.015 | 0.030 | 0.030 |
| $\mu_n$ ($\mu_p$) ($cm^2$/V s) | 32 (8) | 20 (4) | 32 (8) | 32 (8) | 20 (12) | 20 (4) |
| DOS ($cm^{-3}$) (annealed) | $3 \times 10^{18}$ | $3 \times 10^{18}$ | $4 \times 10^{18}$ | $2 \times 10^{18}$ | $7 \times 10^{14}$ | $5 \times 10^{15}$ |
| DOS ($cm^{-3}$) (stabilized) | $3 \times 10^{18}$ | $3 \times 10^{18}$ | $4 \times 10^{18}$ | $2 \times 10^{18}$ | $2 \times 10^{16}$ | $10^{17}$ |
| $\sigma_n$ (tails) ($cm^2$) | $10^{-17}$ | $10^{-17}$ | $10^{-17}$ | $10^{-17}$ | $10^{-17}$ | $10^{-16}$ |
| $\sigma_c$ (tails) ($cm^2$) | $10^{-16}$ | $10^{-16}$ | $10^{-16}$ | $10^{-15}$ | $10^{-15}$ | $5 \times 10^{-16}$ |
| $\sigma_n$ (DB) ($cm^2$) (annealed) | $10^{-16}$ | $10^{-16}$ | $10^{-16}$ | $5 \times 10^{-17}$ | $10^{-15}$ | $10^{-16}$ |
| $\sigma_c$ (DB) ($cm^2$) (annealed) | $5 \times 10^{-16}$ | $5 \times 10^{-16}$ | $5 \times 10^{-16}$ | $5 \times 10^{-15}$ | $5 \times 10^{-15}$ | $5 \times 10^{-16}$ |
| $\sigma_n$ (DB) ($cm^2$) (stabilized) | $10^{-16}$ | $10^{-16}$ | $10^{-16}$ | $5 \times 10^{-17}$ | $10^{-15}$ | $2 \times 10^{-16}$ |
| $\sigma_c$ (DB) ($cm^2$) (stabilized) | $5 \times 10^{-16}$ | $5 \times 10^{-16}$ | $5 \times 10^{-16}$ | $5 \times 10^{-15}$ | $2 \times 10^{-14}$ | $2 \times 10^{-14}$ |

The a-Si:H PIN cells were deposited in the conventional structure: glass/ZnO:Al/P/I/N/ZnO:Al/Ag (or simply Ag back eflector) (Fig. 2a). However it has been shown in a previous article [11] combining modelling with experiments, that a pm-Si:H intrinsic layer cell deposited in this order develops a very damaged P/I interface due to bombardment by the high dose of hydrogen required for developing pm-Si:H. Other experiments conducted in our laboratory [12] further highlight that upon light-soaking treatments, hydrogen bubbles form and lead to macroscopic defects, such as blisters and delamination, that obviously cannot be reversed by annealing. Therefore, as we are interested in the stabilised efficiency, we have modelled pm-Si:H cells with the material deposited as follows: glass/silver/ZnO:Al/N/I/P/ZnO:Al/silver grid in that order (Fig. 2b). The high flux of hydrogen required for pm-Si:H deposition will then damage the N/I interface, to which, as will be shown later by modelling, the solar cell output is much less sensitive. Also macroscopic defects, if any, will now appear at the ZnO:Al/N contact and will therefore have far less repercussion on solar cell characteristics. The light of course should still enter through the P-layer during cell operation for improved efficiency, since the maximum number of holes needs to be created near the contact that is the collector of holes (the ZnO:Al/P contact in this case). This is because, even though the mobility of holes in pm-Si:H is a factor of 3 higher than that in a-Si:H [8, 9], it is still less than the electron mobility. The a-Si:H PIN cell does not have a very damaged P/I interface [11], since this material can be deposited by the decomposition of $SiH_4$ without or very little hydrogen dilution. We therefore commence the investigations in this article on a-Si:H PIN cells, deposited in the normal manner with the P-layer first on ZnO:Al coated glass; but pm-Si:H cells deposited with N-layer first on glass/silver/ZnO:Al. For any deposition involving a rich hydrogen environment, ZnO is preferable to either ITO or $SnO_2$, as ZnO is more stable against reduction by atomic hydrogen. The RMS roughness of the LPCVD grown ZnO:Al is around 60 nm corresponding to a haze of ~23%, measured at a wavelength of 550 nm.

To attain the stabilised state, the samples were light soaked at 100 °C in order to accelerate the kinetics of creation of metastable defects [22]. They were exposed to a maximum intensity of 670 mW $cm^{-2}$ unfiltered white light produced by a 1 kW halogen lamp and then quenched to 30 °C for their characterisation [1]. In some other light-soaking (LS) experiments [19], the cells were light soaked in the open-circuit condition using a mercury vapour lamp providing an intensity of 100 $mW/cm^2$. During LS, the cells were fan-cooled to reduce illumination-induced heating. The temperature of the cells was monitored using a PT100 thermocouple during LS, to verify that the device temperature remained under 50 °C. Light-soaking was continued for 500 h.

P-$\mu$c-Si:H, P-a-SiC:H and P-$\mu$c-$SiO_x$ layers have also been developed in our laboratory with properties given in Table 1. The values given in the table are representative of a series of each such film grown under the same conditions. The CRIND's of the P-layers are compared in Figures 1c and 1d. The conductivity of our P-$\mu$c-$SiO_x$ films is strongly dependent on the $CO_2$flow rate [23], nevertheless, its conductivity compares well with that of optimised P-aSiC:H layers. The properties of the N-$\mu$c-$SiO_x$ have however been taken from reference [17] and its CRIND's are compared to those of standard N-a-Si:H and the TCO ZnO:Al in Figures 1e and 1f.

# 3 SIMULATION MODEL

Our one-dimensional electrical-optical model ASDMP (Amorphous Semiconductor Device Modelling Programme) [24, 25], later extended to model crystalline silicon and HIT cells [26], solves the Poisson's equation and the two carrier continuity equations under steady state conditions for the given device structure and yields the dark and illuminated current density-voltage (*J-V*) characteristics and the QE. The electrical part of the modelling programme is described in references [27, 28]. The expressions for the free and trapped charges, the recombination term, the boundary conditions and the solution technique in ASDMP are similar to the AMPS computer programme [29], developed by Prof. Fonash's group.

The gap state model in ASDMP consists of two monovalent donor-like and acceptor-like tail states and two monovalent Gaussian distribution functions (one being of donor type and the other of acceptor type) to simulate the deep dangling bond (DB) states (as e.g. used in [29] (AMPS), [30, 31] (AFORS-HET)). A more realistic gap state distribution model in a-Si:H would consist of, besides the monovalent band tail states, a deep defect distribution of DBs determined from the defect pool model (DPM) [32, 33]. However in ASDMP, as in AMPS and AFORS-HET, since only monovalent states can be introduced, the deep DB distribution determined from DPM is replaced by two Gaussian distributions of monovalent states, donor like and acceptor like, separated by a correlation energy. Such a replacement has proved to be quite accurate in a-Si:H [34, 35].

**Table 2**. Sensitivity of a pm-Si:H solar cell output to a damaged N/I and a similarly damaged P/I interface (modelling results). The cells given here have a two P-layer structure with light entering the device on the P-layer side.

| ZnO:Al/P-$\mu$c-Si/P-$\mu$c-SiO$_x$/I-pm-Si:H/ N-$\mu$c-SiO$_x$/ZnO:Al/Ag/glass | | $J_{sc}$ (mA cm$^{-2}$) | $V_{oc}$ (volts) | FF | Efficiency (%) |
|---|---|---|---|---|---|
| With a damaged N/I interface | Initial | 15.29 | 1.140 | 0.772 | 13.45 |
| | Stabilised | 15.15 | 1.073 | 0.701 | 11.40 |
| With a damaged P/I interface | Initial | 15.41 | 1.063 | 0.665 | 10.90 |
| | Stabilised | 15.26 | 1.044 | 0.608 | 9.69 |

The contact barrier heights for a cell with the P-layer in contact with a TCO at $x = 0$ and the N-layer in contact with a TCO or metal at $x = L$, are given by: $\varphi_{B0} = E_{\mu}(P) - E_{ac}(P) - sbb$ (1)and $\varphi_{BL} = E_{ac}(N)$. (2)Here $E_{\mu}(P)$ and $E_{ac}(P)$ are the mobility gap and the activation energy of the P-layer, and $E_{ac}(N)$ is the activation energy of the N-layer, forming an ohmic contact at the back of the device. 'sbb' is the TCO/P surface band bending, which we have taken to be ~0.16 eV, the measured value [14] for this quantity for P-$\mu$c-Si:H, employed in all double window structures. A value of 0.2 eV has been assumed for P-$\mu$c-$SiO_x$:H, in simulations involving such a single P-layer. However, later in Section 4 we show the effect of varying the sbb (P-$\mu$c-$SiO_x$:H) on the solar cell output. With the activation energies and mobility band gaps of P-$\mu$c-Si:H and P-$\mu$c-$SiO_x$:H given in Table 1, $\phi_{B0}$ in the present PIN devices is 1.32 eV in the case of TCO/P-$\mu$c-Si:H and 1.83 eV for TCO/P-$\mu$c-$SiO_x$:H. $\phi_{BL}$ is 0.2 eV for the case of N-a-Si:H and 0.09 eV for N-$\mu$c-$SiO_x$:H [19].

The generation term in the continuity equations has been calculated using a semi-empirical model [36] that has been integrated into the modelling programme [24, 25]. Both specular interference effects and diffused reflectances and transmittances due to the interface roughness are taken into account. As mentioned earlier, in the model it is possible to consider only two rough interfaces. The diffused reflection and transmission from the TCO (ZnO:Al) used in the simulations have both a wavelength and angular dependence. In some of the simulations the effect of using a $MgF_2$ ARC is given. This layer is assumed to have a thickness of 50 nm – however it was not optimised by changing its thickness. The complex refractive index for each layer of the structure is required as input and has been measured in-house by spectroscopic ellipsometry. In all cases studied in this article, experimentally or by modelling, light enters through the P-layer, the junction of the TCO with the P-layer being taken as $x = 0$ on the position scale in the modelling calculations. Voltage is also applied at$x = 0$, with the N/TCO or N/metal contact at $x = L$, at the end of the semiconductor layers of the device, being at ground potential.

The input material and device parameters of the model are given in Table 1. The values chosen for a-Si:H are the best average values for the material developed in our laboratory. The parameters for pm-Si:H (e.g., hole mobility, initial and stabilised DOS, etc.) are those deduced from measured data in references [1, 5, 8, 9, 10]. The superior hole transport properties of pm-Si:H, as already pointed out in the introductory section, have been deduced from diffusion induced time resolved microwave conductivity [8] and time-of-flight measurements [9]. The DOS at the Fermi level $N(E_F)$ in the initial state has been measured on pm-Si:H samples using both capacitance measurements on Schottky barriers and space-charge-limited current (SCLC) measurements on $N^+/I/N^+$ structures [5]. From both techniques $N(E_F)$ values of 7–8 × $10^{14}cm^{-3}eV^{-1}$have been obtained which is almost an order of magnitude lower than that measured [5] for a-Si:H (~5 × $10^{15}cm^{-3}eV^{-1}$) or reported in the literature for standard a-Si:H. In reference [1], the evolution of the sub-band gap absorption during accelerated light-soaking at 100 °C was studied for a-Si:H and pm-Si:H films. It was observed that the value of $\alpha(1.1\ eV)$ at "saturation", namely $\alpha_{ss}$, is about an order of magnitude lower for pm-Si:H relative to a-Si:H. All the above results are representative of the study of a large number of samples grown under the same conditions. The Tauc's gap was determined from standard optical transmission measurements [1]. For both a-Si:H and pm-Si:H we have assumed the mobility gap to be~0.1 eV larger than the corresponding optical gap.

# 4 RESULTS AND DISCUSSION

In performing the modelling we have relied heavily on the input parameters derived from simulating a-Si:H and pm-Si:H solar cells described in reference [11]. The reason for using the P-$\mu$c-Si:H/P-a-SiC:H design is given in detail in reference [13] and for using the P-$\mu$c-Si:H/P-$\mu$c-$SiO_x$:H window structure described in the introductory section. However as already stated, we have also examined the P-$\mu$c-$SiO_x$:H single window structure, assuming here a surface band bending of 0.2 eV, as for P-a-SiC:H – a value that is uncertain. The input parameters of the principal layers are given in Table 1 and their complex refractive indices shown in Figure 1. The P/I interface dangling bond DOS is 2 × $10^{16}cm^{-3}$ over 20 nm for the a-Si:H PIN cells, as inferred by modelling experimental results [11]. As the deposition of a pm-Si:H film involves a high flux of hydrogen to the substrate, we have preferred to study solar cells with this material deposited in a NIP configuration with the N-layer first on ZnO:Al/Ag/glass. The improvement in a pm-si:H solar cell output parameters when it is deposited in a NIP configuration has been demonstrated experimentally in reference [19]. This is because solar cell output is less sensitive to a damaged N/I interface than to a similarly damaged I/P interface. This fact is demonstrated in Table 2 by modelling. In actual experiments, for pm-Si:H solar cells deposited in PIN and NIP configurations in our laboratory [19], it has been found that cells deposited in the latter configuration degrade much less – from an initial efficiency of ~9.3% to about 8.2%; while those deposited in the PIN configuration degrade from a lower initial efficiency of ~8.5% to 5.8%, or even down to 3%, in the worst cases after 500 h of light-soaking. The reason for this has been attributed to the creation of macroscopic defects such as blisters and delamination after light-soaking [19]. Coming back to modelling, the N/I interface defect density for the pm-Si:H cells is taken =5 × $10^{17}cm^{-3}$ over the first 5 nm, followed by an intermediate layer with a DOS of 2 × $10^{16}cm^{-3}$ over 30 nm, The existence of such an intermediate layer in pm-Si:H cells has been suggested by modelling in reference [11]. As already stated, in all cases an ultra-thin intrinsic buffer layer follows the window P-layers, both to prevent boron tailing into the main intrinsic layer and to achieve a lower interface defect density. As described in the introduction the back contact was either N-a-Si:H/ZnO:Al/Ag or N-$\mu$c-$SiO_x$:H/Ag with the ZnO:Al left out in the case of a-Si:H solar cells deposited in the PIN configuration, but retaining a thin ZnO:Al layer even for the latter design for the case of the pm-Si:H films, that are deposited in the NIP configuration. As mentioned earlier, this is to prevent silver diffusion into N-$\mu$c-$SiO_x$:H.

## 4.1 The intrinsic layer

The best solar cell output parameters of the a-Si:H solar cells deposited in our laboratory are given in the first row of Table 3and those of the best pm-Si:H solar cells in the first row of Table 4. The remaining rows of the two tables give modelling results with improvements in the output parameters likely to occur when various emitter designs and N-layers are used. For both types of material, we studied the effect of the substrate temperature to see whether any improvement in the absorption coefficient, linked to the extinction coefficients ($\kappa$) could be achieved. No such increase was observed for a-Si:H over a range of temperatures around 200 °C, but the $\kappa$'s of pm-Si:H increased somewhat when the deposition temperature was increased from 175 °C to 210 °C (Fig. 1a), resulting in superior values of $J_{sc}$ (compare the results of the second and third rows of Tab.4) and the EQE given in Figure 4a.

**Table 3** . Solar cell output parameters of a-Si:H PIN cells with different window layers and back contacts in the initial and light-stabilised states. Light enters through the glass/ZnO:Al/P-layer. a-Si:H layer thickness is 200 nm. ZnO in the table corresponds to aluminium-doped ZnO. The first row gives our best experimental results [23] with a 20 nm pm-Si:H P/I buffer, while the rest are model calculations. The last line shows the effect of using $MgF_2$ ARC on glass.

| Cell structure | | $J_{sc}$ (mA $cm^{-2}$) | $V_{oc}$ (volts) | FF | Efficiency (%) |
|---|---|---|---|---|---|
| Expt: Glass/ZnO/P-$\mu c$-$SiO_x$/ | Initial | 15.17 | 0.87 | 0.729 | 9.63 |
| pm-Si:H/I-a-Si:H/N-$\mu c$-Si:H/ZnO/Ag | Stabilised | 14.65 | 0.843 | 0.636 | 7.85 |
| Glass/ZnO/P-$\mu c$-Si/P-a-SiC/ | Initial | 15.55 | 0.988 | 0.782 | 12.01 |
| I-a-Si:H/N-a-Si:H/ZnO/Ag | Stabilised | 15.01 | 0.933 | 0.603 | 8.83 |
| Glass/ZnO/P-$\mu c$-Si/P-$\mu c$-$SiO_x$ | Initial | 15.78 | 1.029 | 0.723 | 11.74 |
| /I-a-Si:H/N-a-Si:H/ZnO/Ag | Stabilised | 15.37 | 0.943 | 0.611 | 8.86 |
| Glass/ZnO/P-$\mu c$-Si/P-$\mu c$-$SiO_x$/ | Initial | 16.07 | 1.032 | 0.729 | 12.09 |
| I-a-Si:H/N-$\mu c$-$SiO_x$/Ag | Stabilised | 15.53 | 0.944 | 0.612 | 8.97 |
| $MgF_2$/Glass/ZnO/P-$\mu c$-Si/ | Initial | 16.29 | 1.032 | 0.729 | 12.25 |
| P-$\mu c$-$SiO_x$ /I-a-Si:H/N-$\mu c$-$SiO_x$ /Ag | Stabilised | 15.74 | 0.944 | 0.612 | 9.09 |

**Table 4** . Solar cell output of pm-Si:H solar cells, deposited in the NIP configuration, but with the light entering on the P-side. The first row gives our best experimental results [19], while the rest are model calculations. In the second row, properties of pm-Si:H deposited at 175 °C are used, while in the rest those of pm-Si:H deposited at 210 °C are used. ZnO in the table corresponds to aluminium-doped ZnO. The subsequent lines show the effect of different window layers and back contacts in the initial and light-stabilised states. The ZnO:Al/P-$\mu$c-SiO$_x$ sbb is taken =0.2 eV. The last two rows show the effect of using $MgF_2$ ARC on ZnO:Al. Thickness of the pm-Si:H layer is 250 nm except for the last case where it is 150 nm.

| Cell type | | $J_{sc}$ (mA cm$^{-2}$) | $V_{oc}$ (volts) | FF | Efficiency (%) |
|---|---|---|---|---|---|
| Expt: ITO/P-$\mu$c-Si/P-$\mu$c-SiO$_x$/ | Initial | 15.7 | 0.95 | 0.63 | 9.3 |
| I-pm-Si:H/N-a-Si:H/SnO$_2$:F/glass | Stabilised | 15.0 | 0.93 | 0.60 | 8.2 |
| ZnO/P-$\mu$c-Si/P-a-SiC/I-pm-Si:H | Initial | 13.31 | 0.985 | 0.792 | 10.39 |
| (175 °C)/N-a-Si:H/ZnO/Ag/glass | Stabilised | 13.22 | 0.980 | 0.744 | 9.63 |
| ZnO/P-$\mu$c-Si/P-a-SiC/I-pm-Si:H | Initial | 14.48 | 0.987 | 0.792 | 11.32 |
| (210 °C)/N-a-Si:H/ZnO/Ag/glass | Stabilised | 14.38 | 0.982 | 0.743 | 10.5 |
| ZnO/P-$\mu$c-Si/P-$\mu$c-SiO$_x$/I-pm-Si:H/ | Initial | 15.02 | 1.106 | 0.748 | 12.43 |
| N-a-Si:H/ZnO/Ag/glass | Stabilised | 14.89 | 1.061 | 0.694 | 10.97 |
| ZnO/P-$\mu$c-Si/P-$\mu$c-SiO$_x$/I-pm-Si:H/ | Initial | 15.29 | 1.140 | 0.772 | 13.45 |
| N-$\mu$c-SiO$_x$/ZnO/Ag/glass | Stabilised | 15.15 | 1.073 | 0.701 | 11.40 |
| ZnO/P-$\mu$c-SiO$_x$/I-pm-Si:H/ | Initial | 15.89 | 1.124 | 0.760 | 13.56 |
| N-$\mu$c-SiO$_x$/ZnO/Ag/glass | Stabilised | 15.74 | 1.057 | 0.687 | 11.43 |
| MgF$_2$/ZnO/P-$\mu$c-SiO$_x$/I-pm-Si:H/ | Initial | 16.64 | 1.125 | 0,759 | 14.20 |
| N-$\mu$c-SiO$_x$/ZnO/Ag/glass | Stabilised | 16.49 | 1.059 | 0.686 | 11.98 |
| MgF$_2$/ZnO/P-$\mu$c-SiO$_x$/ | Initial | 15.98 | 1.129 | 0.764 | 13.78 |
| I-pm-Si:H/N-$\mu$c-SiO$_x$/ZnO/Ag/glass | Stabilised | 15.92 | 1.080 | 0.714 | 12.27 |

Comparing Tables 3 and 4 for the a-Si:H and pm-Si:H PIN cells respectively, we note that although in most cases, the initial $J_{sc}$ in a-Si:H cells is higher, the stabilised efficiency is lower than in pm-Si:H cells. The reason for higher current is the stronger absorption (higher coefficient of extinction – Fig. 1a) of long wavelength visible light in a-Si:H relative to pm-Si:H. On the other hand the stabilised dangling bond density in a-Si:H is one order of magnitude higher than in pm-Si:H (see Tab.1) and this fact in conjunction with a lower hole mobility than in pm-Si:H brings down sharply the stabilised efficiency in a-Si:H PIN cells. In fact, we have been able to achieve 12% stabilised efficiency in pm-Si:H solar cells (Tab. 4), but not as yet in a-Si:H cells (Tab. 3) using parameters extracted by simulating our experimental results.

**Table 5**. Sensitivity of the pm-Si:H solar cell output to the band edge line-up between P-$\mu$c-$SiO_x$:H and pm-Si:H in cells employing a single window layer in the initial state. $\Delta E_c$ and $\Delta E_v$ are respectively the conduction band and valence band discontinuities and $\Delta E_\mu$ the band gap discontinuity between the two. In Table 4 throughout we had assumed $\Delta E_c = \Delta E_v = \Delta E_\mu / 2$.

| ZnO:Al/P-$\mu$c-$SiO_x$/I-pm-Si:H/ N-$\mu$c-$SiO_x$/ZnO:Al/Ag/glass | $V_{bi}$ (volts) | $J_{sc}$ (mA $cm^{-2}$) | $V_{oc}$ (volts) | FF | Efficiency (%) |
|---|---|---|---|---|---|
| P-$\mu$c-$SiO_x$/pm-Si $\Delta E_c = \Delta E_v = \Delta E_\mu/2$ | 1.566 | 15.89 | 1.124 | 0.760 | 13.56 |
| P-$\mu$c-$SiO_x$/pm-Si $\Delta E_c = 0$, $\Delta E_v = \Delta E_\mu = 0.54$ eV | 1.814 | 16.26 | 1.162 | 0.198 | 3.74 |
| P-$\mu$c-$SiO_x$/pm-Si $\Delta E_v = 0$, $\Delta E_c = \Delta E_\mu = 0.54$ eV | 1.314 | 15.82 | 1.143 | 0.736 | 13.31 |

## 4.2 The window and emitter design

Figures 3a and 3b depict respectively typical band diagrams of a-Si:H and pm-Si:H PIN cells having a two-window P-$\mu$c-Si:H/P-$\mu$c-$SiO_x$:H structure and a $\mu$c-$SiO_x$:H N-layer, under AM1.5 light illumination and short-circuit conditions, in the initial state. On the other hand, Figure 4b shows the EQE curves of a pm-Si:H single junction cell with the following doped layers: (i) P-$\mu$c-Si:H/P-a-SiC:H emitter layer and a-Si:H N-layer, (ii) P-$\mu$c-Si:H/P-$\mu$c-$SiO_x$:H and N-a-Si:H, (iii) P-$\mu$c-Si:H/P-$\mu$c-$SiO_x$:H and and N-$\mu$-$SiO_x$:H and (iv) a single P-$\mu$c-$SiO_x$:H emitter with N-$\mu$-$SiO_x$:H. Additionally, Figure 4c shows the effect of a $MgF_2$ anti-reflection coating (ARC) on the last case, while Figure 4d depicts the stabilised state EQE of Figure 4c. In Figure 5 we draw the EQE curves corresponding to the a-Si:H cells described in the second, third and fourth rows of Table 3 in the initial (Fig.5a) and the light-stabilised (Fig. 5b) states. A $MgF_2$ ARC on glass produces negligible improvement as seen by comparing the last two rows of Table 3; hence this case is not included in Figure 5.

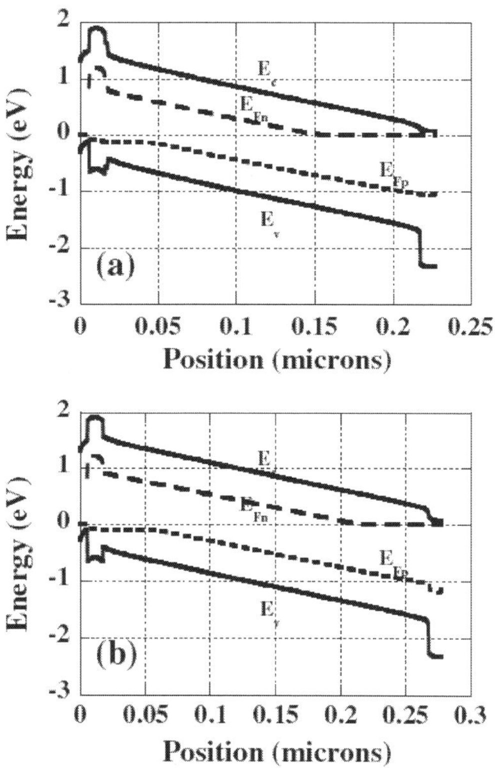


**Figure 3.** Typical band diagrams under AM 1.5 light and short-circuit conditions in the annealed state of (i) a a-Si:H single junction solar cell having a P-μc-Si:H/P-μc-SiOx:H double window design and a μc-SiOx N-layer; and (ii) a pm-Si:H solar cell with identical emitter and N-layer designs under the same conditions.

In the results given in Tables 3 and 4, the P-a-SiC:H and I-a-Si:H or I-pm-Si:H band discontinuity is apportioned to the conduction band side, as inferred from reference [16]. Since there is no valence band discontinuity, the FF is better in this case than when P-$\mu$c-$SiO_x$:H is used. In the latter case, in the absence of any measured data (to our knowledge), we have apportioned half the band discontinuity to the valence band and this fact hinders hole collection and brings down the FF. Also the sbb is 0.16 eV for TCO/P-$\mu$c-Si:H [14]. As there are no experimental data for the TCO/P-$\mu$c-$SiO_x$:H band bending, we have taken the value of 0.2 eV (sixth row of Tab. 4), that is generally encountered at the TCO/P-a-SiC:H contact. The consequences on the solar cell output of changing the band edge line up and TCO/P-$\mu$c-$SiO_x$:H sbb are shown in Tables 5 and6, respectively. Note that the effects of changing the band edge line-up and the sbb for the single P-$\mu$c-$SiO_x$:H design are shown only for the pm-Si:H PIN cell, since the variation is similar in a-Si:H cells, and, such a cell is more promising than an a-Si:H one, from the point of view of the stabilised efficiency. From Table 5, apportioning the entire band gap discontinuity to the valence band side leads to a higher built-in field and $V_{oc}$, but a strongly reduced FF, as the high $\Delta E_v$ hinders hole collection. However, in spite of $\Delta E_c$ being taken equal to zero for this case (a fact that may be expected to encourage electron back diffusion leading to a lower current), the current actually increases, indicating that the high $V_{bi}$, and consequently the higher field in this device, more than cancels the detrimental effect of $\Delta E_c = 0$ on $J_{sc}$. When the entire band gap discontinuity is on the conduction band side (row 3 of Tab. 5), no improvement in $J_{sc}$ is seen over row 1 of the same table, indicating that even $\Delta E_c = \Delta E_\mu/ 2$ is sufficient to block the back diffusion of electrons. $V_{bi}$ decreases for this case as expected, leading to a fall in the FF relative to row 1; surprisingly however $V_{oc}$ increases. This fact may be explained as follows: as $\Delta E_v = 0$, and as we have assumed low I/P (P-layer deposited on top of I layer) interface defect density (our pm-Si:H cell has been deposited in a NIP configuration, so that the I/P interface is not damaged), negligible hole accumulation at this interface takes place for this case, leading to low interface field and more field in the bulk of the device that brings up$V_{oc}$.

Now comparing the results of Veneri et al. [17] to the solar cell output parameters of the a-Si:H cells given in Table 3, we note that our $V_{oc}$'s are considerably higher. Modelling also yields a higher $V_{oc}$ than that achieved experimentally [19, 23]. In such a heterojunction structure, where the P-$\mu$c-$SiO_x$ band gap is considerably higher than that of I-a-Si:H or I-pm-Si:H (Tab.1), the band edge line up has a strong influence on the FF (as seen from the modelling results in Tab. 5 for the case of a pm-Si:H cell, the effect on an a-Si:H cell is similar). Therefore, taking a comparison with the FF values given in reference [17], 19 and 23, no more than half the band gap discontinuity can lie on the valence band side, as assumed by us for all cases in Table 3. However the fact that our calculated $V_{oc}$ is higher than the measured results [17, 19, 23], independently of this band edge line up (as shown in the example of pm-Si:H cells in Tab. 5), indicates that in the experimental cases, the P/I interface is more defective than assumed in our calculations, and points to the fact that the $V_{oc}$ and the cell efficiency may be considerably improved by improving the quality of the P/I interface.

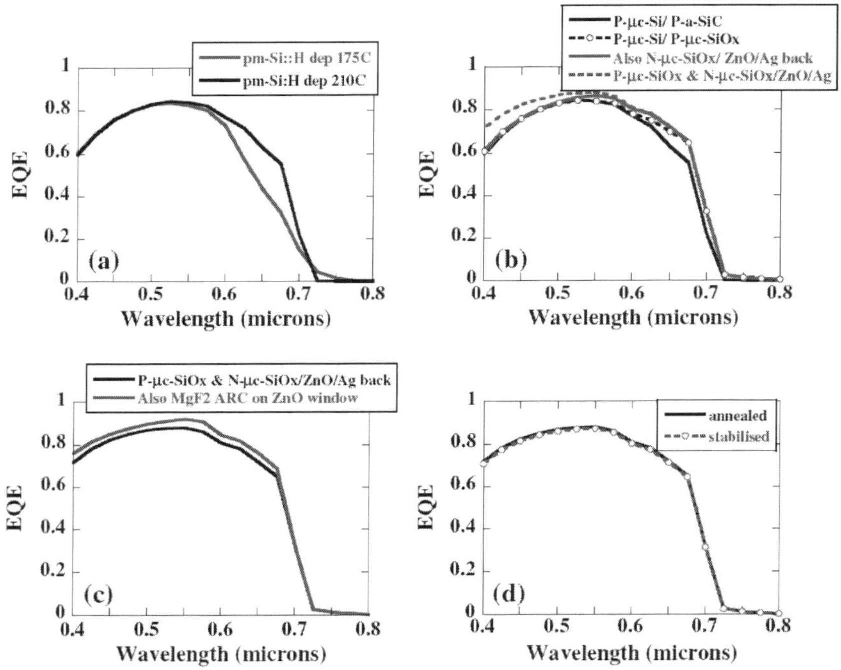


**Figure 4.** Comparison of the external quantum efficiency (EQE) of pm-Si:H solar cells having (a) P-µc-Si:H/P-a-SiC:H double window design with an a-Si:H N-layer, but pm-Si:H intrinsic layers deposited at 175 ◦C and 210 ◦C in the initial state; (b) 4 types of solar cells in the initial state with P-µc-Si:H/P-a-SiC:H and P-µc-Si:H/P-µc-SiOx double window designs both with a-Si:H N-layers, the latter emitter design with N-a-Si:H replaced by N-µc-SiOx and the last design: with a single P-µc-SiOx:H emitter; (c) the latter design in the initial state with and without a MgF2 anti-reflection coating on the ZnO:Al window and (d) the ZnO:Al/P-µc-SiOx/I-pm-Si:H/N-µc-SiOx/ZnO:Al/Ag/glass cell in the initial and light-stabilised states.

**Table 6** . Sensitivity of the pm-Si:H solar cell output to the ZnO:Al/P-$\mu$c-$SiO_x$:H surface band bending. The case where the sbb = 0.2 eV (that is generally observed for ZnO:Al/P-a-SiC:H), is given in the first row of this table and assumed for the single window designs in Table 4. The front contact barrier height ($\phi_{b0}$) for this case is 1.83 eV, since the band gap of P-$\mu$c-$SiO_x$ is taken to be 2.5 eV and its activation energy = 0.47 eV. Even for a slight increase of sbb – to 0.3 eV ($\phi_{b0}$ = 1.73 eV), $V_{oc}$, FF and the cell efficiency deteriorate appreciably (second line of table), as the $V_{bi}$ falls. Any effort to increase the $V_{bi}$ by increasing the P-$\mu$c-$SiO_x$:H thickness, brings down the $J_{sc}$. The third line indicates the effect of having $\phi_{b0}$ = 1.32 eV, the same as at the ZnO:Al/P-$\mu$c-Si:H interface of the double window design.

| Cell type | | $J_{sc}$ (mA cm$^{-2}$) | $V_{oc}$ (volts) | FF | Efficiency (%) |
|---|---|---|---|---|---|
| pm-Si:H, ZnO:Al/P-$\mu$c-$SiO_x$:H sbb = 0.2 eV, $\varphi_{b0}$ = 1.83 eV | Initial | 15.89 | 1.124 | 0.760 | 13.56 |
| pm-Si:H, ZnO:Al/P-$\mu$c-$SiO_x$:H sbb = 0.3 eV, $\varphi_{b0}$ = 1.73 eV | Initial | 15.88 | 1.039 | 0.732 | 12.07 |
| pm-Si:H, ZnO:Al/P-$\mu$c-$SiO_x$:H sbb = 0.71 eV $\varphi_{b0}$ = 1.32 eV | Initial | 15.81 | 0.629 | 0.614 | 6.11 |

**Table 7**. Optimisation of the thickness of the pm-Si:H absorber layer for a double P-layer window design: P-$\mu$c-Si:H/P-$\mu$c-SiO$_x$:H and N-$\mu$c-SiO$_x$/ZnO:Al/Ag back structure. Results indicate that any thickness between 250 nm and 150 nm is good with the thinnest pm-Si:H cell having an edge regarding the stabilised efficiency.

| ZnO:Al/P-$\mu$c-Si/P-$\mu$c-SiO$_x$/I-pm-Si:H/N-$\mu$c-SiO$_x$/ZnO:Al/Ag/glass | | $J_{sc}$ (mA cm$^{-2}$) | $V_{oc}$(volts) | FF | Efficiency (%) |
|---|---|---|---|---|---|
| Thickness = 250 nm | Initial | 15.29 | 1.140 | 0.772 | 13.45 |
| | Stabilised | 15.15 | 1.073 | 0.701 | 11.40 |
| Thickness = 300 nm | Initial | 15.48 | 1.138 | 0.767 | 13.51 |
| | Stabilised | 15.29 | 1.065 | 0.685 | 11.16 |
| Thickness = 200 nm | Initial | 14.99 | 1.143 | 0.775 | 13.29 |
| | Stabilised | 14.90 | 1.082 | 0.715 | 11.53 |
| Thickness = 150 nm | Initial | 14.61 | 1.145 | 0.781 | 13.07 |
| | Stabilised | 14.56 | 1.094 | 0.731 | 11.64 |

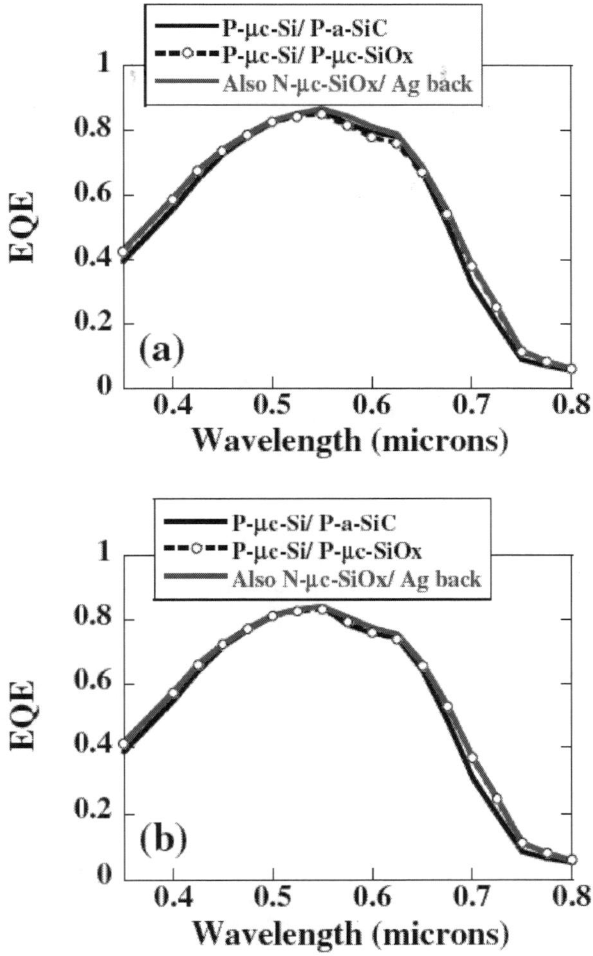


**Figure 5.** Comparison of the EQE curves of a-Si:H solar cells in the (a) initial and (b) light stabilised states having different window designs and N-layer: with P-μc-Si:H/P-a-SiC:H and P-μc-Si:H/P-μc-SiOx:H double window designs both with a-Si:H N-layers and the latter emitter design with N-a-Si:H replaced by N-μc-SiOx.

We find from Tables 3 and 4 and Figures 4b and 5, that in both a-Si:H and pm-Si:H cells there is an increase in the current density when the P-a-SiC:H layer is substituted by P-$\mu$c-$SiO_x$:H. This fact is due to the lower absorption coefficients of P-$\mu$c-$SiO_x$ that is linked to the imaginary part of the CRIND's input to the program – Figure 1c (the real part of the CRIND's of the different P-layers studied are plotted as a function of wavelength in Fig. 1d). The EQE curves indicate that it is mainly the increase in the long wave length response that contributes to the current increase. The absence of an appreciable increase in the short wavelength EQE is due to the fact that P-$\mu$c-SiO$_x$, being micro-crystalline, has been assumed thicker than the amorphous P-a-SiC:H layer, so that a possible increase in the blue response due to lower absorption coefficients of the former is annulled. Our EQE curves in Figure 5a for a PIN a-Si:H cell both having a P-$\mu$c-SiO$_x$ window layer but with the back contact design N-a-Si:H/ZnO:Al/Ag in one case (black dotted line) and N-$\mu$c-SiO$_x$/Ag in the other (red line) agree quite well with the EQE curves of Veneri et al. [17] ($R$ = 0 N-$\mu$c-Si/ZnO/Ag and N-$\mu$c-SiO$_x$/Ag back contacts) except for wavelengths in the range of 0.65 $\mu m \leqslant \lambda \leqslant 0.7 \mu m$, where our EQE curves are lower in both cases. This may be due to lower values of the imaginary part of the CRIND's for those wavelengths for the a-Si:H material developed in our laboratory.

Also from Table 4 and Figure 4b, we note that for the single window design with P-$\mu$c-SiO$_x$:H, the current is appreciably higher, since P-$\mu$c-Si:H is absent, but $V_{oc}$ and FF are lower (because now we cannot take advantage of the lower $E_{ac}$ and sbb of P-$\mu$c-Si:H) yielding nearly the same initial and stabilised efficiency. However in the case of the single P-$\mu$c-SiO$_x$:H window layer the uncertainty of the value of the TCO/P-$\mu$c-SiO$_x$:H sbb may be a crucial limiting factor as is evident from Table 6. Due to the large sensitivity to the TCO/P-$\mu$c-SiO$_x$:H sbb about which little is known at the moment and because, as seen from Table 4, there is negligible difference between the initial and stabilised efficiencies for the double window structure – ZnO:Al/P-$\mu$c-Si:H/P-$\mu$c-SiO$_x$:H, and the single window structure – ZnO:Al/P-$\mu$c-SiO$_x$:H (when this sbb is taken =0.2 eV), it was decided to choose the former design to optimise the thickness of the pm-Si:H film grown at 210 °C. This is because for this case we have a fairly good idea the ZnO:Al/P-$\mu$c-Si:H sbb [14]. The results of thickness optimisation are given in Table 7and show that any thickness between 250 nm and 150 nm is acceptable, with the thinnest pm-Si:H cell having a small advantage regarding the stabilised efficiency.

Finally a $MgF_2$ ARC deposited on ZnO:Al, through which light enters the pmSi:H cell with the NIP configuration on glass (Fig.2b), strongly enhances $J_{sc}$ (Tab. 4) and the EQE (Fig. 4c). However, only a small improvement is observed in the case of the a-Si:H cell in the PIN configuration with $MgF_2$ deposited on glass (Tab. 3).

## 4.3 The N-side design

Tables 3 and 4 and the EQE curves of Figures 4b and 5 show the effect of substituting N-$\mu$c-SiO$_x$:H in place of N-a-Si:H. This brings up $J_{sc}$ and the long wavelength EQE. For this layer the electrical and optical parameters have been taken from reference [17] in our simulation. This material with the real part *n*, of the complex refractive index considerably lower than that of N-a-Si:H and close to that of ZnO:Al (Fig. 1f), appreciably improves the long wavelength EQE. Indeed most of the light is now reflected at the I-layer/N-$\mu$c-SiO$_x$ interface and therefore does not need to cross the N-layer twice. Also it is nearly as transparent as ZnO:Al, with values of the extinction coefficient close to those of the latter (Fig. 1e). Since as well, its doping can be adjusted to make it highly conducting ($E_{ac}$ = 0.09eV, lower than that of N-a-Si:H – Tab. 1), it can effectively act both as the TCO and the N-layer of the structure [17]; also leading to improvements in $V_{oc}$ and FF. Moreover, the deposition of an extra TCO layer may be avoided in this case. However the pm-Si:H cell has to be deposited in a NIP configuration, to avoid a very damaged I/P interface and macroscopic defects as already stated. Here a thin layer of ZnO:Al needs to be deposited on silver, to prevent the latter diffusing into the N-layer when the N-side of the junction is bombarded by a high flux of hydrogen required to form pm-Si:H. Thus in pm-Si:H cells, although the deposition of an extra TCO layer cannot be avoided, the use of N-$\mu$c-SiO$_x$:H instead of N-a-Si:H is still advantageous as can be seen from the 5th row of Table 4, and Figure 4b.

# 5 CONCLUSIONS

The possibility of single junction pm-Si:H solar cells with stabilised efficiencies around 12% has been demonstrated by modelling. We have shown that in order to achieve such efficiencies the pm-Si:H cell should be deposited in the NIP rather than the PIN configuration to avoid a heavily damaged P/I interface that is much more harmful for the solar cell efficiency than a similarly damaged N/I interface; and to avoid formation of macroscopic defects on the P-layer side. However light should enter the cell through the P-layer. A P-$\mu$c-SiO$_x$:H window layer, a suitably N-doped $\mu$c-SiO$_x$:H, as well as a $MgF_2$ ARC on ZnO:Al are required to achieve this goal. On the contrary, using the initial and stabilised state properties and interface parameters of the best a-Si:H material and solar cell developed in our laboratory so far, we could not achieve 10% stabilised efficiency in a-Si:H PIN solar cells, as the stabilised DOS in this material is high and its hole mobility considerably lower than in pm-Si. However such a landmark has been achieved in a few other laboratories e.g. reference [18], probably due to better light absorption in their material, improved P/I interface properties in solar cells fabricated with this material, etc.

The calculated solar cells have a lower FF but a higher $V_{oc}$ when P-$\mu$c-$SiO_x$:H forms part of the window layer than for the P-a-SiC:H case. The higher $V_{oc}$ is due to a higher built-in field in the former case, while the lower FF for P-$\mu$c-$SiO_x$ is due to an assumed $\Delta E_v$ = 0.26eV, that hinders hole collection, while $\Delta E_v$ is ~0 for P-a-SiC:H. The solar cells using P-$\mu$c-$SiO_x$:H show higher current density, due to the lower coefficient of extinction of this material compared to P-a-SiC:H. Also modelling shows that highly conducting and transparent N-$\mu$c-$SiO_x$:H can act both as the N-layer of the structure and the TCO. In fact this material with a refractive index considerably lower than that of N-a-Si:H and close to that of ZnO:Al, appreciably improves the long wavelength QE, since most of the light is now reflected at the I-layer/N-$\mu$c-$SiO_x$ interface and does not need to cross the N-layer twice. The improvements in the solar cell output parameters due to the use of the P- and N-layers described above and as deduced from modelling will be applicable independent of the absorber material used in the PIN solar cell.

In our calculations, there is uncertainty in the calculated solar cell output, due to the limited knowledge on the band edge line up of P-$\mu$c-$SiO_x$ with I-a-Si:H or I-pm-Si:H, and its surface band bending at the contact with ZnO:Al when used in the single window design. The effect of varying these parameters has been demonstrated. The latter uncertainty may be avoided by using the double window design. Also we have shown that $\Delta E_v$ cannot be higher than half the band gap discontinuity in order to attain the high experimental values of FF [17, 19, 23]. Moreover we infer that the $V_{oc}$ in the deposited solar cells may be considerably increased by lowering the P/I interface defect density. We conclude, as Table 4 indicates, that ~12% stabilised efficiency is achievable in thin pm-Si:H single junction solar cells (thickness of the I-layer ⩽250 nm) with practical material and device parameters and a suitable anti-reflection coating.

## ACKNOWLEDGMENTS

This work was carried out in the framework of the FP7 project "Fast Track", funded by the EC under grant agreement No. 283501. The computer modelling code was developed (electrical part) by P. Chatterjee during projects funded by MNRE and DST, Government of India, and the optical part was added during her tenure as Marie Curie fellow at Ecole Polytechnique, Palaiseau, France.

## REFERENCES


1. R. Butté, R. Meaudre, M. Meaudre, S. Vignoli, C. Longeaud, J.P. Kleider, P. Roca i Cabarrocas, Phil. Mag. B **79**, 1079 (1999)
2. P.P. Ray, P. Choudhuri, P. Chatterjee, Thin Solid Films **403–4**, 275 (2002)
3. Baojie Yan, Guozhen Yue, L. Sivec, J. Yang, Subhendu Guha, C.-S. Jiang, Appl. Phys. Lett. **99**, 113512 (2011)
4. P. Roca i Cabarrocas, A. Fontcuberta i Morral, S. Lebib, Y. Poissant, Pure Appl. Chem. **74**, 359 (2002)
5. J.P. Kleider, C. Longeaud, M. Gauthier, M. Meaudre, R. Meaudre, R. Butté, S. Vignoli, P. Roca i Cabarrocas, Appl. Phys. Lett. **75**, 3351 (1999)
6. P. St'ahel, S. Hamma, P. Sladek, P. Roca i Cabarrocas, J. Non-Cryst. Solids **227–230**, 276 (1998)
7. P. Roca i Cabarrocas, Mater. Res. Soc. Symp. Proc. **507**, 855 (1998)
8. A. Fontcuberta i Morral, R. Brenot, E.A.G. Hamers, R. Vanderhaghen, P. Roca i Cabarrocas, J. Non-Cryst. Solids **266–269**, 48 (2000)
9. E.A. Schiff, J. Non-Cryst. Solids **352**, 1087 (2006)
10. M. Brinza, G. Adriaenssens, P. Roca i Cabarrocas, Thin Solid Films **427**, 123 (2003)
11. Y. Poissant, P. Chatterjee, P. Roca i Cabarrocas, J. Appl. Phys. **94**, 7305 (2003)
12. K.-H. Kim, E.V. Johnson, A. Abramov, P. Roca i Cabarrocas, Sol. Energy Mater. Sol. Cells **105**, 208 (2012)
13. N. Palit, P. Chatterjee, J. Appl. Phys. **86**, 6879 (1999)

14. W. Ma, T. Saida, C.C. Lim, S. Aoyama, H. Okamoto, Y. Hamakawa, in *Proceedings of the First World Conference on Photovoltaic Solar Energy Conversion, Hawaii* (IEEE, New York, 1994), p. 117

15. X. Xu, J. Yang, A. Banerjee, S. Guha, K. Vasanth, S. Wagner, Appl. Phys. Lett. **67**, 2323 (1995)

16. Z.Y. Wu, J.M. Siefert, B. Equer, in *Proc. 10th EU PVSEC, Lisbon, Portugal, 1991*, p. 953.

17. P.D. Veneri, L.V. Mercaldo, I. Usatii, Appl. Phys. Lett. **97**, 023512 (2010)

18. A. Lambertz, F. Finger, R.E.I. Schropp, U. Rau, V. Smirnov, Prog. Photovolt.: Res. Appl. **23**, 939 (2015)

19. K.-H. Kim, S. Kasouit, E.V. Johnson, P. Roca i Cabarrocas, Sol. Energy Mater. Sol. Cells **119**, 124 (2013)

20. P. Roca i Cabarrocas, J.B. Chévrier, J. Huc, A. Lloret, J.Y. Parey, J.P.M. Schmitt, J. Vac. Sci. Technol. A **9**, 2331 (1991)

21. Y.M. Soro, A. Abramov, M.E. Gueunier-Ferrat, E.V. Johnson, C. Longeaud, P. Roca i Cabarrocas, J.P. Kleider, J. Non-Cryst. Solids **354**, 2092 (2008)

22. S. Vignoli, R. Meaudre, M. Meaudre, Phil. Mag. B **73**, 261 (1996)

23. S.N. Abolmasov, H. Woo, R. Planques, J. Holovský, E.V. Johnson, A. Purkrt, P. Roca i Cabarrocas, Eur. Phys. J. Photovolt. **5**, 55206 (2014)

24. P. Chatterjee, M. Favre, F. Leblanc, J. Perrin, Mater. Res. Soc. Symp. Proc. **426**, 593 (1996)

25. N. Palit, P. Chatterjee, Sol. Energy Mater. Sol. Cells **53**, 235 (1998)

26. Madhumita Nath, P. Chatterjee, J. Damon-Lacoste, P. Roca i Cabarrocas, J. Appl. Phys. **103**, 034506 (2008)

27. P. Chatterjee, J. Appl. Phys. **76**, 1301 (1994)

28. P. Chatterjee, J. Appl. Phys. **79**, 7339 (1996)

29. P.J. McElheny, J.K. Arch, H.-S. Lin, S.J. Fonash, J. Appl. Phys. **64**, 1254 (1988)

30. R. Stangl, M. Kriegel, K. v. Maydell, L. Korte, M. Schmidt, W. Fuhs, in *Conference record 31st IEEE Photovoltaic Specialists Conference, 2005*, p. 1556

31. R. Varache, J.P. Kleider, W. Favre, L. Korte, J. Appl. Phys. **112**, 123717 (2012)

32. M.J. Powell, S.C. Deane, Phys. Rev. B **48**, 10815 (1993)

33. M.J. Powell, S.C. Deane, Phys. Rev. B **53**, 10121 (1996)

34. V. Halpern, Phil. Mag. B **54**, 473 (1986)

35. C. Longeaud, J.P. Kleider, Phys. Rev. B **48**, 8715 (1993)

36. F. Leblanc, J. Perrin, J. Schmitt, J. Appl. Phys. **75**, 1074 (1994)

# Chapter 5

# Enhancing Solar Cell Efficiency Using Photon Upconversion Materials

**Yunfei Shang [1], Shuwei Hao [1,2], Chunhui Yang [1,2,*] and Guanying Chen [1,3,*]**

[1]School of Chemical Engineering and Technology, Harbin Institute of Technology, Harbin 150001, China

[2]Harbin Huigong Technology Co., Ltd., Harbin 150001, China

[3]Institute for Lasers, Photonics, and Biophotonics, University at Buffalo, State University of New York, Buffalo, NY 14260, USA

## ABSTRACT

Photovoltaic cells are able to convert sunlight into electricity, providing enough of the most abundant and cleanest energy to cover our energy needs. However, the efficiency of current photovoltaics is significantly impeded by the transmission loss of sub-band-gap photons. Photon upconversion is a promising route to circumvent this problem by converting these transmitted sub-band-gap photons into above-band-gap light, where solar cells typically have high quantum efficiency. Here, we summarize recent progress on varying types of efficient upconversion materials as well as their outstanding uses in a series of solar cells, including silicon solar cells (crystalline and amorphous), gallium arsenide (GaAs) solar cells, dye-sensitized solar cells, and other types of solar cells. The challenge and prospect of upconversion materials for photovoltaic applications are also discussed.

## KEYWORDS

photovoltaic; upconversion; efficiency

## 1. INTRODUCTION

Fossil fuels (coal, oil, and natural gas) form the major energy source for meeting current human needs [1,2,3,4], yet cause a range of serious environmental issues. Moreover, the ever-growing consumption rate outpaces their regeneration rate, endangering the exhaustion of fossil fuels on earth. Dealing with the energy crisis is an urgent need [5,6]. Among all new competing energy sources (biomass, wind, hydroelectricity, geothermal energy, and nuclear energy), solar energy is considered to be the most abundant, renewable, and environment-friendly energy form [7,8]. The total solar power that strikes the Earth's surface is about 100,000 terawatts, which is 10,000 times more than that consumed globally [9,10]. If just 0.1% of the sunlight that reaches the Earth's surface could be converted by photovoltaic (PV) devices, with an average conversion efficiency of 10%, the amount would be sufficient to meet our current energy demands. The achievement of PV technology in recent decades has been tremendous [11,12,13]. The high-cost per kilowatt delivered by PV stations, however, limits its competitiveness with other sources. This is caused mainly by the inferior power conversion efficiencies of PV devices. The inability to absorb infrared (IR) light (700–2500 nm), which constitutes 52% of the energy of the entire solar spectrum, forms the major energy loss mechanism of conventional solar cells. This fundamental issue is set by the sizable bandgap of light-absorbing materials in PV devices. Crystalline silicon (c-Si) photovoltaic (PV) cells are the most used among all types of solar cells on the market, representing about 90% of the world's total PV cell production in 2008 [14]. However, even for single crystalline silicon (Si) PV cells with a rather small semiconductor band-gap (1.12 eV, corresponding to a wavelength of ~1100 nm), the transmission loss of sub-band-gap photons can still amount to about 20% of the sun's energy irradiated onto the Earth's surface [15]. For PV cells with a larger band-gap, such as amorphous Si (1.75 eV) solar cells, which are limited to absorb sunlight with wavelengths below 708 nm, they manifest even higher near infrared transmission losses.

Photon upconversion (UC) provides a means to circumvent transmission loss by converting two sub-band-gap photons into one above-band-gap photon, where the PV cell has high light responsivity. This technology enables us to break the Shockley–Queisser limit of a single-junction PV cell (about 31% for non-concentrated sunlight irradiation for a semiconductor material with an optimized band-gap of around 1.35 eV) by transforming the solar spectrum [16]. Indeed, Trupke *et al.* demonstrated through a detailed balance model that modification of the solar spectrum with an up-converter could elevate the upper theoretical efficiency limit of a single-junction crystalline silicon PV cell to be as high as 40.2% under non-concentrated sunlight irradiation [17]. This value is far beyond the Shockley–Queisser limit for crystalline silicon solar cells (~1.1 eV bandgap) of approximately 30%. Figure 1 schematically illustrates the use of upconversion processes to convert the solar spectrum in the IR-Near IR (NIR)-short visible range into the peak (~500 nm) of sun radiation. There are three typical photon upconversion materials under investigation now outlined. (1) Rare-earth-doped micro- and nano-crystals (RED-UC), which usually work with wavelengths above 800 nm, but also possibly below 800 nm with appropriate composition design [18,19,20,21,22,23,24]. The abundant electronic states of trivalent lanthanide ions enable upconverting a range of IR wavelengths by selecting varied types of rare earth ions [25,26]. (2) Triplet–triplet annihilation upconversion (TTA-UC, response range $\lambda < 800$ nm) [27,28,29], whereby the triplet states of two organic molecules interact with each other, exciting one molecule to its emitting state to produce fluorescence. The excitation power density required for TTA UC is quite low, a few mW/cm$^2$, which is comparable to that of sun radiation. (3) Upconversion in Quantum Nanostructures (QN-UC, response range $\lambda < 800$ nm) [30,31,32]. This process leans on the use of a unique design comprising a compound semiconductor nanocrystal, which incorporates two quantum dots with different bandgaps separated by a tunneling barrier. Upconversion occurs by excitation of an electron in the lower energy transition, followed by intra-band absorption of the hole, allowing it to cross the barrier to a higher energy state. Detailed mechanisms of these three types of upconversion materials are discussed in Section 2. These upconversion materials are now emerging in use to improve PV efficiency in the real word.

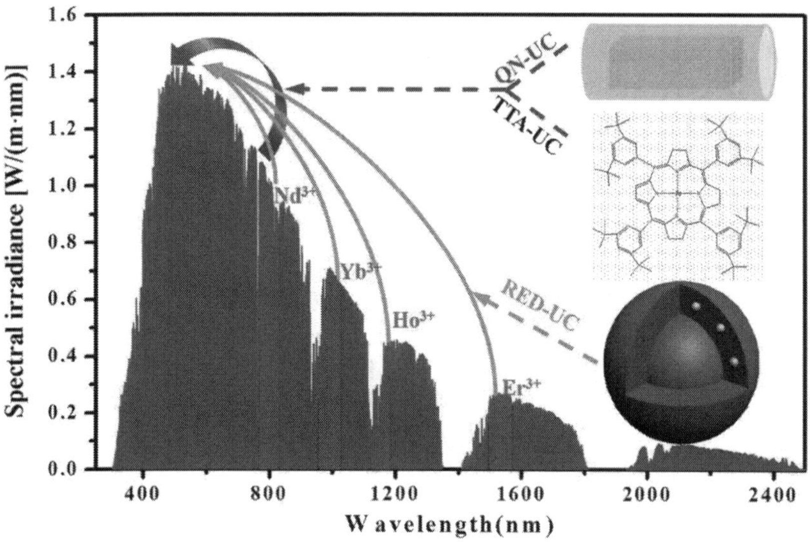


**Figure 1.** The absorption and emission range of three types of upconversion materials in reference to AM 1.5 spectrum (QN-UC (purple): upconversion in quantum nanostructures; TTA-UC (purple): triplet-triplet annihilation upconversion; RED-UC (green): Rare-earth-doped upconversion materials).

# 2. UPCONVERSION MATERIALS

## 2.1. Rare-Earth-Doped Upconversion Materials

The rare-earth family is comprised of 17 elements, which includes 15 lanthanide elements (from La to Lu) plus the elements of yttrium (Y) and scandium (Sc). The trivalent lanthanide ions possess a $4f^n5s^25p^6$ electronic structure with 14 available orbitals ($0 < n < 14$), offering 14 possible electronic group configurations. The quantum interaction of involved electrons endows lanthanide elements with abundant energy levels covering a spectral range of NIR, visible and ultraviolet (UV) [33,34,35, 36,37]. In addition, the perfect shielding of $4f$ electrons by outer complete $5s$ and $5p$ shells enables electronic transitions to occur with limited influence from the surrounding environment, thus exhibiting high resistance to processes of photobleaching and photochemical degradation. As the symmetries of involved quantum states are identical, the intra-$4f$electronic transitions of lanthanide ions are electric-dipole forbidden, yet can be relaxed due to local-crystal-field-induced intermixing of the $f$ states with higher electronic configurations. The primary forbidden nature yields metastable energy levels of lanthanide ions (lifetime can be as long as tens of milliseconds), thus favoring the occurrence of sequential excitations in excited states of a single lanthanide ion as well as permitting favorable ion-ion interactions in excited states to allow energy transfers between two or more lanthanide ions.

There are four main basic mechanisms for rare-earth ions based upconversion processes, comprising of excited state absorption (ESA), energy transfer upconversion (ETU), photon avalanche (PA) and energy migration-mediated upconversion (EMU) (Figure 2). Excited state absorption (ESA) takes the form of successive absorption of pump photons by a single ion utilizing the ladder-like structure of a simple multi-level system as portrayed in Figure 2a. Figure 2a illustrates a simplified three level system for two sequential photon absorption processes. The ETU process involves at least two types of ions, namely a sensitizer and an activator. In this process, ion I known as the sensitizer is firstly excited from the ground state to its metastable level by absorbing a pump photon; it then successively transfers its harvested energy to the ground state and the first excited state of ion II, known as the activator, exciting ion II to its upper emitting state, which is followed by radiative decay to its ground state. The PA in Figure 2c is a looping process that involves an efficient cross relaxation mechanism between ion I in the ground state and ion II in the second excited state, resulting in generation of two ion IIs in the metastable state. The population of ion II in the second excited state is created through absorption of laser photons at its metastable state (the first excited state), which is initially populated through non-resonant weak ground state absorption. When the looping process ensues, an avalanche population of ion II will be created at its metastable state, producing avalanche upconverted luminescence from the emitting state. The generation of PA UC typically occurs above a certain threshold of excitation density. Below the threshold, very little up-converted fluorescence is produced, while the luminescence intensity increases by orders of magnitude above the pump threshold. In addition, the looping nature enables the evoked UC luminescence to be strongly dependent on the laser pump power, especially around the threshold laser power. The lanthanide ions designed for realizing energy migration-mediated upconversion (EMU) comprise four types: the sensitizers (type I), the accumulators (type II), the migrators (type III), and the activators (type IV; Figure 2d). A sensitizer ion is used to harvest excitation photons and subsequently promotes a neighboring accumulator ion to its excited states. A migrator ion extracts the excitation energy from high-lying energy states of the accumulator, followed by random energy hopping through the migrator ion sublattice and trapping of the migrating energy by an activator ion that produces luminescence by decaying to the ground state. The excitation density for RED-UC is typically in the range of $10^{-1}$–$10^{2}$W/cm$^{2}$.

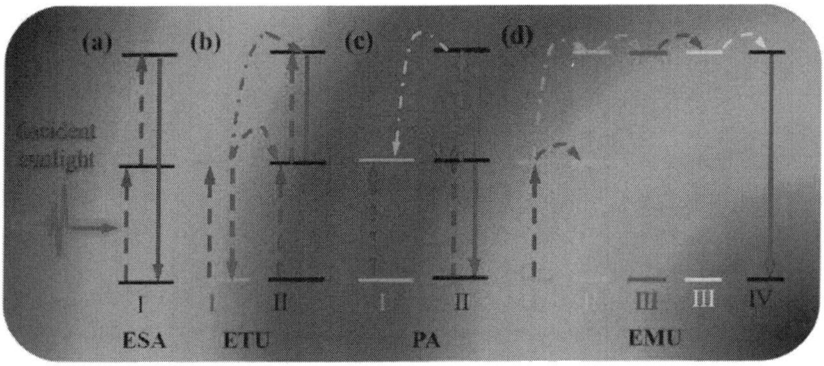


**Figure 2.** A schematic illustration of four typical upconversion processes: (**a**) Excited state absorption (ESA); (**b**) Energy transfer upconversion (ETU); (**c**) Photon avalanche (PA); (**d**) Energy migration-mediated upconversion (EMU).

Rare-earth-doped upconversion materials typically consist of an appropriate dielectric host matrix and doped $Ln^{3+}$ ions that are dispersed as the guest in the lattice of the host matrix. Host materials with low phonon energy are able to produce upconversion luminescence at high efficiency, as multiphonon-assisted nonradiative relaxations between the closely spaced energy levels can be minimized, thus yielding increased lifetime of intermediate energy levels. Investigated low phonon energy host materials typically include fluorides, chlorides, iodides, and bromides [38], while high phonon energy host materials such as silicates, borates, and phosphates are also under study [39]. In general, host materials with low phonon energy are hygroscopic, while the high phonon energy ones are robust even under acute environment (strong acid, base, high temperature, *etc.*). Yet, the type of fluoride host material is unique and has attracted a lot of attentions in recent years. This is because fluoride host lattice not only has low phonon energy but also shows excellent chemical stability. In particular, hexagonal $NaYF_4$ lattice is considered to be one of the most efficient host materials to date [40,41,42]. Interestingly, even for cubic phase $NaYF_4$, a well-defined distribution of $Na^+$ and $Y^{3+}$ ions in the crystal lattice can enable ultrahigh upconversion luminescence [43].

The $Ln^{3+}$ dopants provide light harvesting ability as well as upconverting ability for RED UC. Among $Ln^{3+}$ ions, the research for enhancing the efficiency of solar cells explores the use of single $Ln^{3+}$ doping such as $Er^{3+}$, $Ho^{3+}$, $Tm^{3+}$, and $Pr^{3+}$to upconvert IR light. Meanwhile, a utilization of $Yb^{3+}$ ions as co-dopants can provide new, strong absorption at ~980 nm ($^2F_{7/2} \rightarrow {}^2F_{5/2}$). The $Yb^{3+}$ ions are able to sensitize most lanthanide activator ions, typically, $Er^{3+}$, $Ho^{3+}$, $Tm^{3+}$, resulting in intense upconversion when excited. Recently, $Nd^{3+}$ ion has been proposed as another extraordinary sensitizer. It provides a new absorption at 808 nm and is able to sensitize the $Yb^{3+}$ ion. This further broadens the absorption range of RED upconversion for PV application. Table 1 summarizes some selected upconverting materials which can be excited, utilizing light with wavelength longer than 800 nm. (1) Single $Er^{3+}$ doped UC materials can convert light at 1523 nm to green (550 nm) and red (650 nm) (Figure 3a) [44,45,46,47]; (2) $Yb^{3+}/Er^{3+}$ codoped system can produce green (525 nm, 542 nm), red (655 nm), as well as purple (415 nm, weak) emissions under 980 nm laser excitation [42,48,49,50,51,52,53]; (3) $Yb^{3+}/Tm^{3+}$ pairs are able to convert light at 980 nm into UV (345 nm), blue (480 nm) and NIR (800 nm) emissions [54,55,56,57,58]; (4) $Nd^{3+}/Yb^{3+}/Ln^{3+}$ (Ln = Er, Tm, Ho, *etc.*) tri-doped UC materials can convert 808 nm NIR light to visible luminescence [57,58,59,60,61,62]. As an example, the energy transfer processes for $Nd^{3+}$, $Yb^{3+}$, and $Er^{3+}$ are depicted in Figure 3b for illustration. The $Er^{3+}$ ion emits in the green and red range after absorbing the excitation energy by either the $Nd^{3+}$ or the $Yb^{3+}$ ions. Simultaneous use of $Nd^{3+}$, $Yb^{3+}$ and $Er^{3+}$ ions enables light harvesting at ~800 nm, ~980 nm, as well as ~1523 nm, covering broader spectral range for upconversion.

## 2.2. Triplet-Triplet Annihilation Upconversion Materials

Triplet-triplet annihilation (TTA) up-conversion is a process occurring in a pair of sensitizer-annihilator dyes, as portrayed in Figure 4. Light harvesting is enabled by the sensitizer, followed by populating its singlet excited state ($S_0 \rightarrow S_1$). The intersystem crossing process (ISC, $S_1 \rightarrow T_1$) allows the triplet excited state of the sensitizer (organo-metallic type) to be efficiently populated thanks to the heavy atom effect of the transition metal atom. Subsequently, a transfer of the energy in the triplet to a neighboring acceptor (annihilator) at $S_0$ state *via* a Dexter energy transfer (DET) process, can excite the acceptor (annihilator) to its triplet state. Two nearby acceptor (annihilator) molecules in the triplet states collide with each other, and results in one acceptor molecule being excited to the higher singlet state ($S_1$), while the other one returns to the ground singlet state ($S_0$). A radiative decay from the generated singlet excited state of the acceptor produces an upconverted fluorescence, which is called triplet-triplet annihilation (TTA) upconversion. Furthermore, the wavelength of light harvesting can be varied by selecting the sensitizer, while the emission wavelength of TTA up-conversion can be tuned by selection of the triplet acceptor (annihilator). Since sun radiation can be utilized in a direct way to induce TTA UC, this type of upconversion is promising for applications in PV solar cells. Yet, the inability to produce efficient TTA UC in long visible as well as in the IR range, limits its impact on PV technology. To address this challenge, Kimizuka *et al.* [73] and Bardeen *et al.* [74] extended the absorption wavelength beyond 850 nm by using metallonaphthalocyanines and semiconductor nanocrystals as triplet sensitizers. However, the upconversion quantum efficiency of TTA-UC remains rather low. To address the low efficiency problem, F. Meinardi and co-workers [75] combined organic TTA with fluorescent semiconductor nanocrystals that not only broaden the absorption but also lower the excitation intensity. The same group [76] synthesized a series of sensitizers with tuned absorption and achieved a conversion efficiency of 10% under broadband AM 1.5 irradiation. Kimizuka *et al.* [77] demonstrated a novel metal-organic framework, in which donor-acceptor position could be precisely tailored to achieve a high UC efficiency even under a weak excitation (less than that of sunlight).

TTA UC is highly sensitive to oxygen molecules, as the generated triplet states of TTA dye pairs can be easily quenched by them, producing reactive oxygen species (ROS). These ROS species are exceptionally reactive, which can cause the chemical degradation of the paired dyes. As a result, TTA upconversion will be deactivated by oxygen molecules, posing another problem for their PV applications. To address this issue, Kimizuka and co-workers prepared a solvent-free liquid TTA UC system that can function well even under aerated conditions [78]. Subsequently, they reported a series of air-stable TTA upconversion in supramolecular organogel [79], supramolecular self-assemblies [80], and ionic liquids [81], providing a new approach to solve the problem of the sensitivity to oxygen molecules. Alternatively, Weder *et al.* demonstrated TTA upconversion in molecular glasses and organogels where oxygen-induced upconversion fluorescence quenching could also be prevented [82,83].

**Table 1.** Some selected upconverters excitable with wavelength longer than 800 nm.

| Dopant Ion | Host Material | Excitation (nm) | Emission (nm) | References |
|---|---|---|---|---|
| $Er^{3+}$ | $NaYF_4$ | 1523 | 550, 660, 800, 980 | [63] |
| $Yb^{3+}$-$Er^{3+}$ | $NaYbF_4$ | 980 | 520, 540, 654 | [64] |
| $Yb^{3+}$-$Er^{3+}$ | $NaYF_4$ | 980 | 410, 522, 540, 650 | [65–67] |
| $Yb^{3+}$-$Er^{3+}$ | $NaYF_4$@$NaYF_4$ | 980 | 510~570, 640~680 | [68] |
| $Yb^{3+}$-$Er^{3+}$ | $NaYF_4$@$NaYF_4$:$Nd^{3+}$ | 808/980 | 520, 540, 655 | [69] |
| $Yb^{3+}$-$Tm^{3+}$ | $NaYF_4$ | 980 | 375, 450, 475, 679, 800 | [70,71] |
| $Yb^{3+}$-$Ho^{3+}$ | $NaYF_4$ | 980 | 545, 650 | [72] |
| $Yb^{3+}$-$Er^{3+}$-$Nd^{3+}$ | $NaYF_4$ | 808/980 | 410, 520, 545, 650 | [60] |

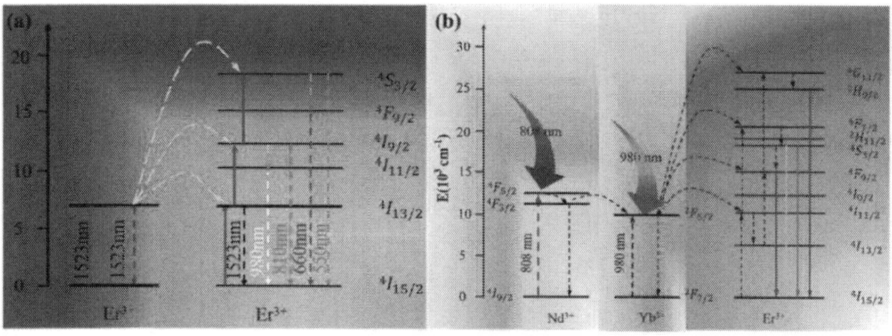


**Figure 3.** The upconversion mechanisms for (**a**) single $Er^{3+}$ doped upconversion particles (UCNPs) excited under 1523 nm laser exciation; (**b**) $Nd^{3+}$-$Yb^{3+}$-$Er^{3+}$ tridoped system under 808 nm or 980 nm laser exciation.

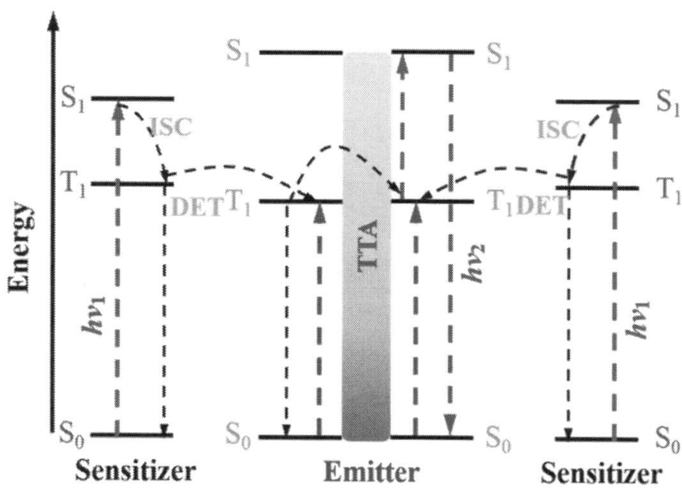


**Figure 4.** The working principle of the triplet-triplet annihilation upconversion.

## 2.3. Upconversion in Quantum Nanostructures

Upconversion in quantum nanostructures is realized through a design of a compound semiconductor nanocrystal, which incorporates two quantum dots with different bandgaps separated and connected by a tunneling barrier (a semiconductor rod). The implementation of upconversion is through a two-step absorption of two subsequent photons. The first photon generates an electron-hole pair via interband absorption in the lower-energy core (small band-gap dots), leaving a confined hole and a delocalized electron in the compound semiconductor nanocrystal. The second absorbed photon can lead, either directly or indirectly, to further excitation of the hole, enabling it to cross the barrier layer. The direct way is through an intraband absorption of the photon by the confined hole at the lower-energy core (Figure 5a), while the indirect way is via an Auger mediated energy transfer process (Figure 5b).The Auger process describes a recombination of the second electron-hole pair, generated by absorbing the second photon, while simultaneously allowing nonradiative energy transfer to the confined hole at the lower energy core (small band-gap dots), empowering it to cross the barrier to the higher energy quantum dot (large band-gap dots). This, in turn, is followed by a radiative recombination with the delocalized electron, producing upconverted luminescence. This system combines the stability of an inorganic crystalline structure, with the spectral tunability afforded by quantum confinement. Since the absorption, emission, and lifetime of the semiconductor nanocrystals can be controlled by variation of their size, shape, as well as composition, upconversion in quantum nanostructures holds prime promise for applications in PV devices. However, a relative high excitation density (~$10^4$W/cm$^2$) is needed to activate this type of upconversion. Lowering the excitation density to the range of sun irradiation (~$10^{-1}$W/cm$^2$) is an inviting direction, which could produce a pronounced impact for PV applications in a straightforward way.

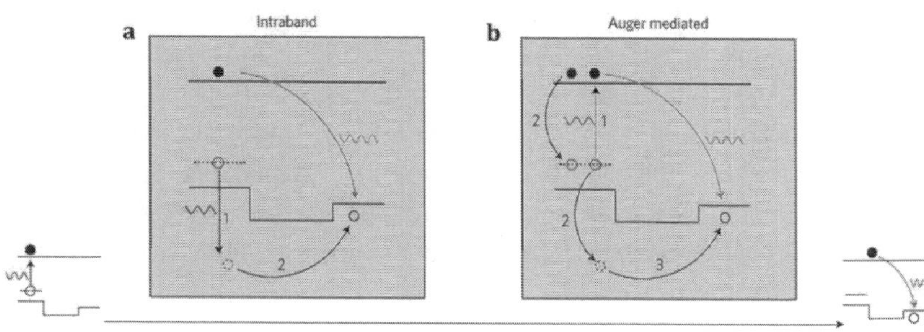


**Figure 5.** Mechanism of the upconversion in quantum nanostructures: (**a**) Direct intraband hole absorption mechanism of upconversion; (**b**) Auger-mediated upconversion (adapted with permission from [30]; Copyright Nature Publishing Group, 2013).

# 3. PV APPLICATIONS

The solar cell, which directly converts solar energy into electricity, is one of the most attractive solutions to the growing energy demands, due to the omnipresence and abundance of solar energy. Light harvesting is the first but also the most important step, which determines how much sun irradiation can be absorbed. The standard solar irradiation spectrum (AM 1.5) covers the wavelength region from UV to IR (300–2500 nm). However, most single-junction PV devices are unable to absorb light quanta in the long wavelength range, typically NIR and IR, creating the performance-limiting issue of transmission loss. Upconversion materials are in play to circumvent this issue by spectral conversion, which has been applied to a wide range of solar cells. These PV devices include c-Si, amorphous Si thin-film, GaAs and dye sensitized solar cells (DSSCs). Two typical configurations have been exploited for applications in PV cells, as portrayed in Figure 6, to upconvert sub-bandgap light into above-bandgap luminescence. Structure One involves the use of a reflection layer to increase the optical path of NIR or IR light within the upconverting layer for an increase of the upconversion luminescence output, while Structure Two does not use any reflection layer, typically employed by DSSCs.

## 3.1. c-Si Solar Cells

Crystalline silicon (c-Si) photovoltaic cells are used in the largest quantity of all types of solar cells on the market, representing about 90% of the world total PV cell production in 2008. The recorded maximum conversion efficiency of crystalline silicon solar has reached 25% [84], close to the Schockley–Queisser limit (~30%). However, as mentioned before, the bandgap of crystalline silicon (~1.12 eV) limits it to absorb light less than 1100 nm, producing a transmission loss as high as 20%. On the other hand, c-Si solar cells work with the highest quantum efficiency in the spectral region of 800–1100 nm. Development of upconversion materials with absorption below 1100 nm, and the upconverted emission in the range of ~800–1100 nm are highly attractive for improvement of the efficiency of a c-Si PV cell. $Yb^{3+}$-sensitized or $Nd^{3+}$-sensitized RED UC materials are beyond consideration for use in c-Si PV cells, as they provide absorption at ~980 nm and ~800 nm where the semiconductor silicon has existing high spectral responses.

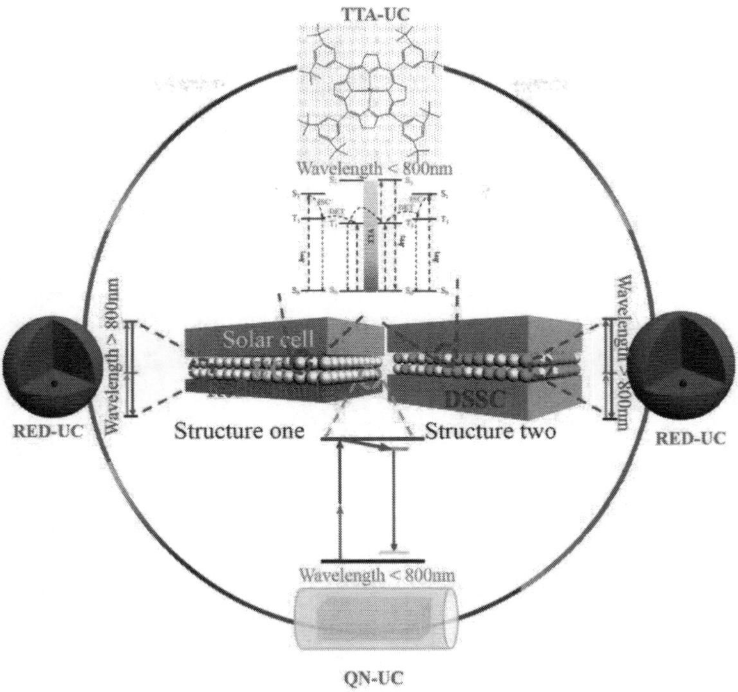


**Figure 6.** A schematic illustration of the usage of three typical upconversion materials in solar cells using two configurations. (Rare-earth upconversion, RED-UC; Upconversion in quantum nanostructure, QN-UC; Triplet-triplet annihilation upconversion, TTA-UC).

Single $Er^{3+}$-doped RED UC materials are attractive for c-Si solar cells, as they display absorption at 145–1580 nm with sub-bandgap energy of silicon, and emit strong luminescence at ~980 nm, ~540 nm, and 650 nm that can be very useful for silicon to produce excitons. In 2005, Shalav *et al.* [63] exploited the use of $NaYF_4$:20% $Er^{3+}$ microcrystals as an upconverter in bifacial c-Si solar cells. In their geometric arrangement of implementation of upconversion, microsized $NaYF_4$:20% $Er^{3+}$phosphors were dispersed into an acrylic adhesive medium with matched refractive index, and then this mixture was deposited as a thick film on the rear of a bifacial c-Si solar cell. An impressive increase of external quantum efficiency of 2.5% was accomplished under excitation at 1523 nm. Fischer *et al.* [47] also investigated the potential use of microsized or bulk $NaYF_4$:20% $Er^{3+}$ materials (~3 μm) to improve the conversion performance of c-Si solar cells. The upconversion efficiency of $NaYF_4$:20% $Er^{3+}$ microcrystals was quantified to be 5.1% when irradiated with 1523 nm laser with a power density of 1880 W/m$^2$. Shalav *et al.* showed that a c-Si solar cell device can have an external quantum efficiency of 0.34% under sub-bandgap irradiation at 1522 nm with a power density of 1090 W/m$^2$.

Alongside $Er^{3+}$ ions, the $Ho^{3+}$ ions provide a possibility to harvest sub-bandgap energy of silicon at a new wavelength range (1150–1230 nm). The centroids of upconverted luminescence of $Ho^{3+}$ ions are at ~910 nm (corresponding to the $^5I_5 \rightarrow ^5I_8$ transition) in the NIR range, and at ~650 nm (corresponding to the $^5F_5 \rightarrow ^5I_8$ transition) in the visible range. Both radiations have energy above the bandgap of silicon. It should be noted that the intensity of sun radiation at 1170 nm is approximately twice than that at 1520 nm, potentially delivering more effective improvement of solar cell efficiency. Lahoz *et al.* [85] first reported on the use of single $Ho^{3+}$ ion doped upconverting glass ceramics in a c-Si solar cell. They successfully demonstrated that the c-Si solar cell responds to light irradiation at ~1170 nm due to the upconverted visible emission (650 nm) and NIR emission (910 nm) in the glass ceramics. In another work of theirs, $Ho^{3+}$-$Yb^{3+}$ codoped upconversion materials were investigated to improve the conversion efficiency of a c-Si solar cell [86]. They found that the $Ho^{3+}$-$Yb^{3+}$ codoped upconverter produced much stronger NIR emission intensity than that of the single $Ho^{3+}$-doped counterpart. This is because the excited $Ho^{3+}$ ions (at the $^5I_5$ state) can sensitize $Yb^{3+}$ ions (at the ground state $^2F_{7/2}$ state), making both $Ho^{3+}$ and $Yb^{3+}$luminescence. This, in turn, result in an improved NIR response of Si solar cells in comparison to the one using single $Ho^{3+}$-doped glass ceramics. Moreover, Lahoz and co-workers [86] validated the combined use of single-$Er^{3+}$ doped RED UC material and single $Ho^{3+}$ doped oxyfluoride RED UC material to encompass broader sunlight harvesting in the NIR range (Figure 7). This concept was demonstrated by placing the $Er^{3+}$ doped RED UC layer to the back of the $Ho^{3+}$ doped RED UC layer, which is then attached to the back of a c-Si solar cell. To enhance NIR light harvesting, a mirror was placed at the bottom of the cell to reflect the unabsorbed sub-bandgap sunlight back to the upconverting layers.

From a material development point of view, Chen *et al.* [87] prepared a core-shell-shell structure of $NaGdF_4{:}Er^{3+}@NaGdF_4{:}Ho^{3+}@NaGdF_4$ nanocrystals, in which the $Er^{3+}$ and the $Ho^{3+}$ ions were separately doped into the core and the first shell layer (Figure 8). Intense UC emissions from both $Er^{3+}$ and $Ho^{3+}$ were shown in the same core-shell-shell nanoparticle. The intensities of luminescence bands from both ions are all much stronger than that from $Er^{3+}$/$Ho^{3+}$-codoped nanocrystals due to a spatial isolation of $Er^{3+}$ and $Ho^{3+}$ ions, which can avoid the detrimental cross relaxation between these ions. However, because of high doping concentration of $Er^{3+}$ and $Ho^{3+}$, adverse relaxations could still induce quenching beyond the interfaces, even though the dopants were in different layers. To circumvent this, a multi-layer core/shell design of $NaYF_4$:10%$Er^{3+}$@$NaYF_4$@$NaYF_4$:10% $Ho^{3+}$@$NaYF_4$@$NaYF_4$:1% $Tm^{3+}$@$NaYF_4$ nanoparticles were reported by us [88], in which the inert $NaYF_4$ layers in between upconverting domains were utilized to efficiently suppress the detrimental cross-relaxation processes between different types of lanthanide ions, yielding about two times more efficient upconversion photoluminescence than the counterpart $NaYF_4$:10% $Er^{3+}$@$NaYF_4$:10% $Ho^{3+}$@$NaYF_4$:1% $Tm^{3+}$@$NaYF_4$ without the inert $NaYF_4$ layers. Moreover, these core/multishell nanoparticles can be excited at ~1120–1190 nm (due to $Ho^{3+}$), ~1190–1260 nm (due to $Tm^{3+}$), and ~1450–1580 nm (due to $Er^{3+}$), collectively covering a broad spectral range of ~270 nm in the infrared range that can be very useful for infrared photosensitization of c-Si solar cells.

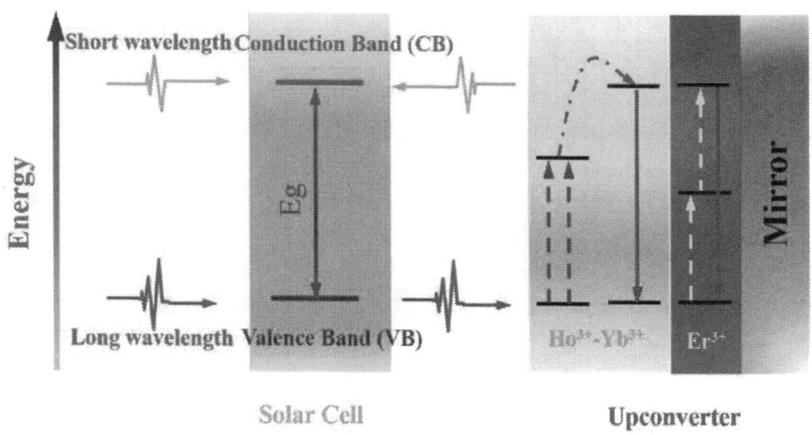


**Figure 7.** An operating mechanism for a c-Si solar cell with two upconverters, one co-doped with $Ho^{3+}$–$Yb^{3+}$ and the other one single doped with $Er^{3+}$. Photons with short wavelength can be absorbed directly by the solar cell. The transmitted sub-bandgap light can be upconverted into high-energy photons, which would be absorbed by c-Si. The mirror behind the upconverter increases the probability of absorption of sub-bandgap light in the upconverter layer.

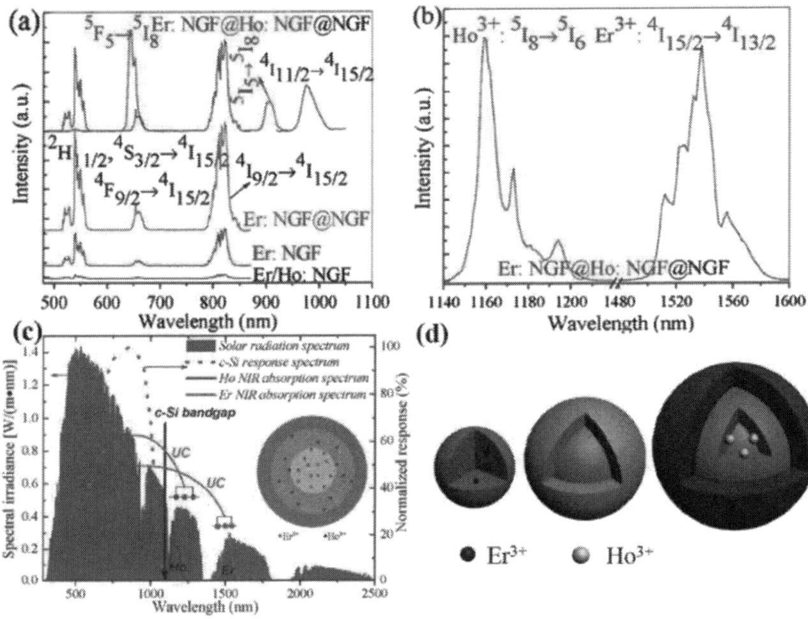


**Figure 8.** (**a**) Upconversion emission spectra of the core $NaGdF_4:Er^{3+}/Ho^{3+}$nanocrystals, the core $NaGdF_4:Er^{3+}$ nanocrystals, the core/shell $NaGdF_4:Er^{3+}$@$NaGdF_4$nanocrystals, and the $NaGdF_4:Er^{3+}$ @$NaGdF_4:Ho^{3+}$@$NaGdF_4$ core/shell/shell nanocrystals; (**b**) Upconversion excitation spectrum of the $NaGdF_4:Er^{3+}$@$NaGdF_4:Ho^{3+}$@$NaGdF_4$ core/shell/shell nanocrystals; (**c**) A schematic illustration of the upconversion luminescence of $NaGdF_4:Er^{3+}$@$NaGdF_4:Ho^{3+}$ @$NaGdF_4$ nanocrystals *versus* the bandgap of c-for Si solar cell, and the spectrum of AM 1.5 sun irradiation; (**d**) A schematic of illustration of different nanostructures: the $NaGdF_4:Er^{3+}$core, the $NaGdF_4:Er^{3+}$@$NaGdF_4$ core/shell, and the $NaGdF_4:Er^{3+}$@$NaGdF_4:Ho^{3+}$@$NaGdF_4$ core/shell/shell (adapted with permission from [87]; Copyright Royal Society of Chemistry, 2012).

In addition, Alexander Dobrovolsky and co-workers [89] designed a system based on GaNP nanowires that can harvest IR light through a two-step two-photon absorption process and emits visible light. Though it is claimed to be suitable for application in third-generation Si-based solar cells, no demonstrations have yet been shown.

## 3.2. Amorphous Silicon Solar Cells

Amorphous Si solar cells have been considered as a promising substitute of c-Si cells due to their low cost, easy preparation, and excellent chemical stability. The band-gap of amorphous Si is larger than that of c-Si at about 1.75 eV, which confines it to absorb NIR light shorter than 700 nm. This means that upconverting materials with absorption above 700 nm should be appealing for uses in amorphous Si solar cells to circumvent the transmission loss.

Table 2 lists a range of investigated and selected single/co-doped RED UC materials in the literature that can be utilized towards this purpose. Zhang *et al.* [90] investigated the use of $NaYF_4$:18% $Yb^{3+}$, 2% $Er^{3+}$ nanocrystals as upconverter in an amorphous Si solar cell. This kind of upconverter shows visible emissions at around 655 nm (red), 525 nm and 540 nm (green) after absorbing light at 980 nm. They showed that the short circuit density can be increased by 6.25% (from 16 to 17 $mA \cdot cm^{-2}$) due to the contribution of these upconverters. In analogy, De Wild *et al.* [91,92] utilized β-$NaYF_4$:$Yb^{3+}$/$Er^{3+}$ as upconverter to enhance the power conversion efficiency of an amorphous Si solar cell. The upconverter layer with a thickness of 200–300 μm was deposited at the rear of the amorphous Si solar cell after mixing with polymethylmethacrylate. The maximum photocurrent was increased from 2.1 to 6.2 μA, along with a manifestation of an external quantum efficiency of 0.02% at 980 nm. $Yb^{3+}$/$Er^{3+}$-codoped $TeO_2$-$PbF_2$ oxyfluoride tellurite glass [93] as well as $Gd_2(MoO_4)_3$:$Er^{3+}$/$Yb^{3+}$ UC materials [94] have also been investigated in a similar way by applying them at the back of amorphous silicon cells. The broader NIR light harvesting ability, due to the effect of host matrix, was supposed to yield a significant improvement in the solar efficiency. Unfortunately, there was only a tiny improvement obtained when co-excited by AM 1.5 and 980 nm laser radiation. The underlying mechanism remains unclear, and deserves further investigations.

In addition, $Er^{3+}$ doped β-$NaYF_4$ powders were also applied to amorphous Si solar cells by Chen and co-workers [100]. This solar cell displayed a current of 0.3 μA and 0.01 μA when irradiated with a laser at 980 nm (power density of 60 $mW/cm^2$) and at 1560 nm (power density of 80 $mW/cm^2$), respectively. An enhanced current of 0.54 μA in solar cells was observed when irradiated with 980 nm (60 $mW/cm^2$) and 1560 nm (100 $mW/cm^2$) simultaneously. This was associated with a broadened NIR response due to simultaneous absorption at both ~980 nm and ~1560 nm. UC materials with broad band NIR absorption are more effective to circumvent transmission losses. In this way, TTA upconversion is in play which exploits the strong and broad absorption of dye molecules. Recently, a TTA-upconverter was investigated in hydrogenated amorphous silicon (a-Si:H) solar cells, producing a high efficiency of 10.1% with an absorption threshold of 700 nm [84]. Cheng and co-workers carried out a great deal of research work on optimizing the design of amorphous silicon solar cells combined with TTA-UC materials [101,102,103,104,105,106,107]. For example, an amorphous silicon solar cell was shown to achieve an obvious increase of current of $2.40 \times 10^{-3} \cdot mA/cm^2$, on equipping with a hybrid-emitter TTA system (namely, a filled cuvette with Ag-coated glass beads) as a back-reflecting medium. The use of TTA-UC to increase the light harvest efficiency of amorphous Si solar cells is still underway.

## 3.3. GaAs Solar Cells

Gibart *et al.* in 1995 firstly reported an application of RED UC material on a GaAs solar cell, demonstrating that it is useful to enhance the solar cell efficiency by improving the harvest of unavailable IR light [108]. In their subsequent experiment [108], a substrate-free GaAs solar cell (a band-gap of 1.43 eV) was coupled to a 100 μm thick vitroceramic upconverting materials which contained $Yb^{3+}$ and $Er^{3+}$ ions. The efficiency of the GaAs solar cell was increased quadratically with the power of excitation due to the nonlinear nature of the upconversion process. A power conversion efficiency as high as 2.5% was achieved under 891 nm (1.391 eV) illumination with an irradiance of 25.6 $W/cm^2$. In a similar way, Lin *et al.* [109] adhered a 300 μm thick UC phosphor layer of $Y_{5.86}W_2O_{15}:0.05Yb^{3+}$, $0.09Er^{3+}$ to the rear of a GaAs solar cell. The maximum output power of $0.339 \times 10^{-6}$ W was obtained with a 973 nm laser irradiation at 145.65 $W/cm^2$. These results clearly indicate that GaAs solar cells can work effectively under sub-bandgap light irradiance. However, the involved light irradiances are much higher than that of sun radiance ($10^{-2}$–$10^{-1}$ $W/cm^2$). Development of efficient UC materials under low irradiance is required.

**Table 2.** Typical dopant ions, the main emissions, and the corresponding energy transitions for upconversion nanocrystals reported in the literature.

| Dopant Ion | Host Material | Excitation (nm) | Emission (nm) | Reference |
|---|---|---|---|---|
| $Er^{3+}$ | $NaYF_4$ | 1523 | 550, 660, 800, 980 | [63] |
| $Er^{3+}$ | $YF_3$ | 1490 | 410, 530, 550, 660, 810, 980 | [95] |
| $Er^{3+}$ | $CaF_2$ | 1510 | 410, 550, 660, 980 | [96] |
| $Er^{3+}$ | $Y_2O_3$ | 1538 | 562, 659, 801, 987 | [97] |
| $Er^{3+}$ | Fluoride glasses | 1538 | 550, 660, 820, 980 | [98] |
| $Er^{3+}$ | $BaCl_2$ | 1535 | 410, 550, 660, 980 | [99] |
| $Ho^{3+}$ | Glass ceramics containing $PbF_2$ nanocrystals | 1170 | 650, 910 | [85] |

## 3.4. Dye-Sensitized Solar Cells

Dye-sensitized solar cells (DSSCs) are third-generation PV cells, which have brought revolutionary innovation to PV technology since the first report by Grätzel in 1991 [110]. Owing to their low cost, simple fabrication methodology, environmental friendliness, as well as flexible structure, DSSCs have been considered as the most promising alternative for Si-based solar cells [111]. However, due to the limited absorption range of investigated dyes, the improvement of the conversion efficiency of a DSSC is always challenging [112,113,114,115]. The band-gap of typically used dyes in DSSC, such as N3, N719, and N749, *etc.*, is usually higher than 1.8 eV. This means that they are only able to absorb photons with wavelengths shorter than 700 nm, leading to unharvesting of approximately 52% of the solar energy in the IR range (from 700 to 2500 nm) [116,117,118]. Efforts to develop panchromatic sensitizers for DSSCs have been hampered by poor electron injection efficiency and competing charge recombination when the absorption spectrum of the sensitizer is extended to the IR region, adversely affecting the efficiency. Moreover, the photostability of an IR absorbing dye is known to be very poor [119,120]. Photon upconversion provides an alternative approach by converting unavailable IR photons into high-energy photons that can be absorbed by the sensitizing dyes with a typical structure as presented in Figure 9.

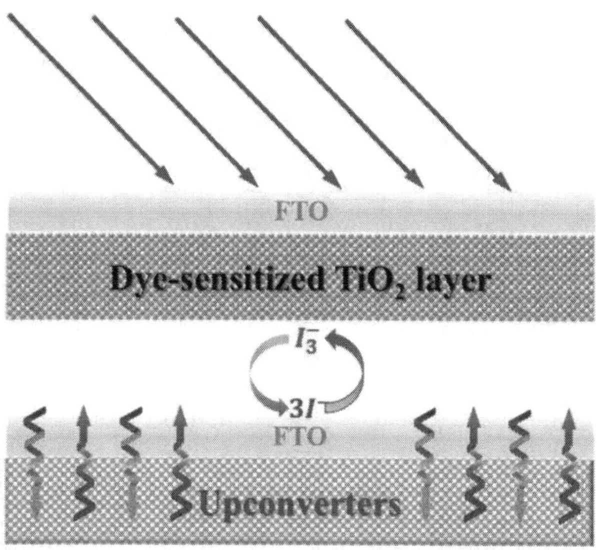


**Figure 9.** A typical schematic configuration of a DSSC equipped with upconverters.

The first DSSC device, based on rare-earth-doped UC materials, was presented by Demopoulos *et al.*, in which $LaF_3:Yb^{3+}/Er^{3+}$-$TiO_2$ nanocomposites were used to form a multilayer electrode structure [121]. As a proof of concept demonstration, the response of DSSC at NIR light (~980 nm) was shown. Later, they deposited a layer of microsized β-$NaYF_4:Yb^{3+}/Er^{3+}$ particles on the rear side of a counter electrode [122], which was able to provide both light reflection and NIR light harvesting without apparent charge recombination at interfaces. An improvement of NIR response was accomplished. In parallel, Yuan *et al.* [123] in 2012 proposed and validated the use of colloidal β-$NaYF_4$:2% $Er^{3+}$/20% $Yb^{3+}$nanoparticles in a DSSC. As the size is less than 20 nm, they are allowed to diffuse in the mesoporous $TiO_2$ layer of the DSSC, enabling them to efficiently interact with sensitization dyes. The advantage of this approach is that these nanoparticles can be utilized in the way of using sensitizing dyes without modifications of device preparation procedure. Following that, Yang and co-workers attached $YF_3:Yb^{3+}/Er^{3+}$ particles to the surface of the porous $TiO_2$ thin film, the power conversion efficiency was increased from 5.18% (blank DSSC without RED UC materials) to 6.76% [124].

Researchers also placed a lot of effort in the application of RED-UC structures in photoanode layers. Wang *et al.* [125] achieved an efficiency increase of 23% when doping $YOF:Yb^{3+}/Er^{3+}$ particles into the $TiO_2$ photoanode layer. Zhang *et al.*exploited the use of $NaYF_4:Er^{3+}/Yb^{3+}$@$TiO_2$ core-shell composite as a photoanode. The conversion efficiency was enhanced by a factor of 1.23 [126]. To avoid electron recombination losses, Zhao *et al.* separated the upconversion and $TiO_2$ layers by growing the middle $SiO_2$ layer between them, achieving an improvement of 29.4% in efficiency [127]. In another way, Demopoulos *et al.* [128] investigated the utilization of β-$NaYF_4:Yb^{3+}/Er^{3+}$@$TiO_2$ submicro particles as both a light harvesting and an IR energy relay layer. An optimization of the layer of the DSSC structure resulted in a 16% relative increase of power conversion efficiency. On the other hand, there are a number of studies using a range of RED UC materials doped with $Yb^{3+}$-$Er^{3+}$ [68,129,130], $Yb^{3+}$-$Tm^{3+}$ [29], $Er^{3+}$ [131], for improving the NIR response of DSSCs, and the following illustration (Figure 10) presents the energy transfer mechanisms in a DSSC equipped with a upconverter codoped with $Yb^{3+}$ and $Er^{3+}$.

Narrow absorption and high excitation threshold of RED-UC are two main limits for efficiency improvement in DSSCs. To mitigate this, Bai and co-workers [132] enhanced light harvesting by 14% due to the usage of $CeO_2:Er^{3+}/Yb^{3+}$ nanofibers that exploited host effect to broaden the light absorption ability. Chen *et al.* [133] provided a method to broaden the NIR absorption of upconversion nanoparticles by utilization of a hexagonal core-shell structure of β-$NaYbF_4$:2% $Er^{3+}$@$NaYF_4$:30% $Nd^{3+}$. This allows simultaneous use of the absorption of $Nd^{3+}$ (~800 nm), $Yb^{3+}$ (~980 nm), and possibly $Er^{3+}$(~1550 nm) ions for light harvesting, while the core/shell structure enables a spatial isolation of $Nd^{3+}$ from $Yb^{3+}$ and $Er^{3+}$ions to avoid detrimental cross relaxations to entail a high upconversion efficiency. Zhao *et al.* [134] achieved an efficiency of 8.32% (with a noticeable enhancement of 14.78%) by putting core-shell-structured β-$NaYF_4$:$Yb^{3+}$/$Er^{3+}$@$SiO_2$@Au nanocomposites on top of a mesoporous $TiO_2$ layer. Yang's group [124,135,136] also did a lot research on enhancing the performance of DSSCs using different lanthanide ion doped nanoparticles. For example, the design of $Ho^{3+}$-$Yb^{3+}$-$F^-$ tri-doped $TiO_2$ nanoparticles enabled a 37% improvement in the power conversion efficiency. Despite recent achievements, improvement of DSSCs efficiency is still limited by the upconversion efficiency as well as the narrow and low absorption ability of RED-UC. Overcoming these two problems would lead to an epidemic use of RED-UC in DSSCs.

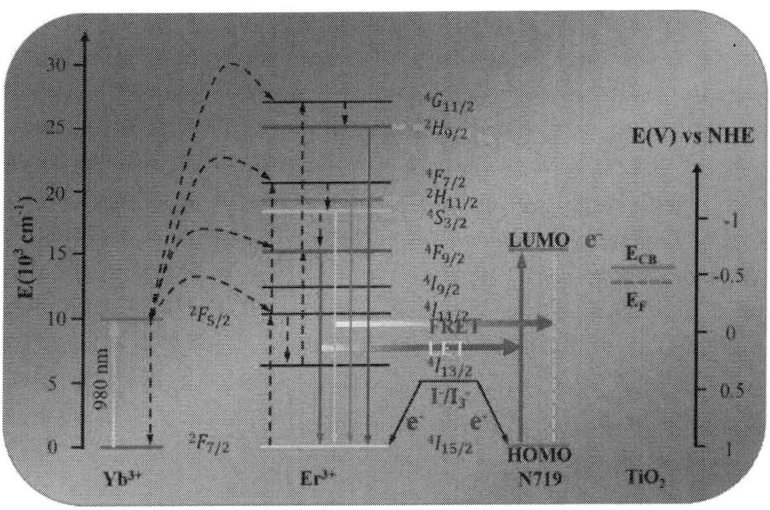


**Figure 10.** A cartoon illustration of energy transfer mechanisms in upconversion nanoparticles (UCNPs)-Doped DSSCs. This figure shows the absorption and conversion of near infrared (NIR) photons into visible light of higher energy via upconversion; the electron transfer from UCNPs to the conduction band of $TiO_2$, as well as the electron transfer from $I^-/I_3^-$ to the ground state of UCNPs. FRET and LET are fluorescence resonance energy transfer and luminescence-mediated energy transfer, respectively. HOMO and LUMO are highest occupied molecular orbital and lowest unoccupied molecular orbital, respectively.

## 3.5. Other Types of Solar Cells

Apart from the solar cells discussed above, organic solar cells are considered to be one of the most promising alternatives for Si-based solar cells, due to their advantages of being flexible, low-cost, light-weight, of simple fabrication and large-scale production for the PV industry [137,138,139,140,141]. However, limited by the spectral mismatch of the absorption of organic molecules with sun irradiation, the improvement in efficiency remains a daunting task. Bulk heterojunction-based PV cells with a large bandgap of organic molecules are only able to absorb visible sunlight. Currently, some organic solar cells have been designed to harvest 800–900 nm sunlight by using low-band-gap polymer materials such as PCPDTBT (poly-[*N*-9′-heptadecanyl-2,7-carbazole-alt-5,5(4′,7′-di-2-thienyl-2′,1′,3′-benzothiadia-zole]) and its derivatives [18,142,143,144]. The poly (3-hexylthiophene) (P3HT) and the fullerenederivative [6,6]-phenyl-C61-butyric acid methyl ester (PCBM) have been exploited to increase the optical response of organic solar cells to 650–700 nm [145].

Upconversion materials have been dedicated to further extend the NIR spectral response. Wang and co-workers utilized $LaF_3$:$Yb^{3+}$/$Er^{3+}$ phosphors to improve the NIR response of P3HT:PCBM organic solar cells. An upconverted photocurrent density of ~16.5 μA/cm$^2$ was obtained under an excitation density of 25 mW/cm$^2$ at 975 nm [146]. In another work of theirs, they demonstrated that the UC material ($MoO_3$:$Yb^{3+}$/$Er^{3+}$), incorporated into P3HT:PCBM organic solar cells, contributed about 1% to the improvement of short-circuit current under one-sun (AM 1.5) illumination [147]. Wu *et al.* [148] in 2012 added $NaYF_4$:$Yb^{3+}$, $Er^{3+}$ nanoparticles on the rear of a P3HT:PCBM organic solar cell, the short-circuit current was enhanced by 0.5 μA when illuminated by a 980 nm light.

To utilize the NIR solar spectrum more efficiently, Adikaari *et al.* [149] reported an application of $Y_2BaZnO_5$:$Yb^{3+}$, $Ho^{3+}$UC particles in PCDTBT:PCBM organic solar cells. Two different layout designs are presented in Figure 11. The PCDTBT:PCBM active layer absorbs photons of wavelength shorter than 700 nm, while $Y_2BaZnO_5$:$Yb^{3+}$/$Ho^{3+}$ shows an intense absorption in the NIR region of 870–1030 nm due to the $^2F_{7/2} \rightarrow$ $^2F_{5/2}$ ($Yb^{3+}$) transition. Moreover, the UC emission peak at 545 nm from $Ho^{3+}$, corresponding to the $^5S_2/^5F_4 \rightarrow {}^5I_8$ transition, matches well with the absorption band of PCDTBT:PCBM active layer. Illuminated by a 986 nm laser with an excitation density of ~390 mW/cm$^2$, a maximum photocurrent density of 16 μA/cm$^2$ and a conversion of 0.45% were obtained, showing the ability to expand the response into the NIR range. Guo *et al.* [150] used a similar method to improve the performance of both NIR harvesting and light scattering of inverted polymer BHI solar cells through mixing $NaYF_4$:$Yb^{3+}$/$Er^{3+}$ with PCDTBT:PCBM.

Instead of using UC materials as upconverters, the porphyrin-based TTA-sensitizers have also been utilized to efficiently increase the NIR response of solar cells [151]. Schulze's group carried out a comprehensive study to explore the application of TTA-UC in organic solar cells [104,106], among which the photocurrent was increased up to 0.2% under a moderate concentration of 19 suns [105]. In this system, a TTA-UC unit was conjugated to the inverted organic cell, to avoid parasitic optical losses [102].

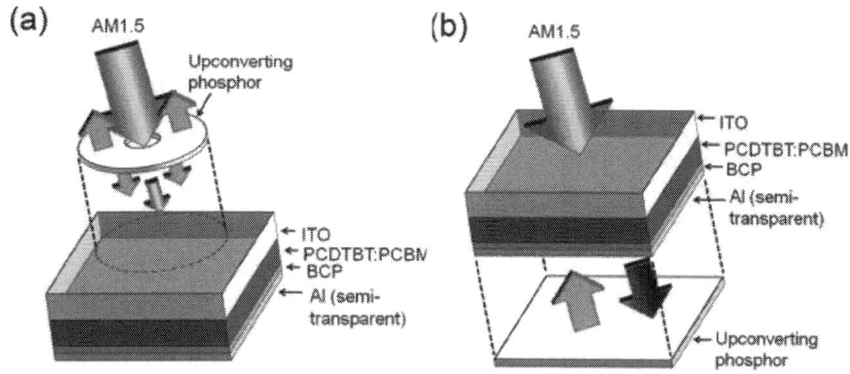


**Figure 11.** Schematic design of an organic PV device with upconversion phosphors placed (**a**) in front of or (**b**) behind the device (adapted with permission from [149]; Copyright AIP Publishing, 2012).

# 4. CONCLUSIONS

In this review, we have summarized photon upconversion materials including rare-earth-doped upconversion materials, triplet-triplet annihilation upconversion materials, and quantum nanostructure upconversion materials as spectral modifiers for various types of photovoltaic solar cells to circumvent their major energy loss mechanism of transmission loss. The ability to convert the transmitted sub-band-gap photons into above-band-gap light where solar cells typically have high quantum efficiency, enables the Schockley-Queisser limit of single junction PV devices to be broken. The inorganic rare-earth-doped upconversion materials are able to work above 800 nm, while the organic triplet-triplet annihilation upconversion dye pairs typically work below 700 nm. Moreover, quantum nanostructures emerge as a new type of upconversion materials, where the size- and shape-induced quantum effects can be exploited to upconvert at therequired wavelength ranges. Indeed, the use of these upconversion materials has improved the performance of c-Si, amorphous Si, GaAs, DSSCs as well as other PV devices with varying bandgaps. Despite recent progress, the increase of efficiency remains less than 2%, far below the theoretical predication of ~10% as in the case of c-Si PV cells when irradiated with unconcentrated sun light [17]. This discrepancy is ascribed to several problems of current photon upconversion materials. (i) The established energy conversion efficiency of long wavelength rare-earth-doped upconversion materials is less than 3% [152], thus limiting the contribution of spectral conversion of absorbed photons for improvement of PV efficiency. To improve the upconversion efficiency, various approaches can be utilized such as non-luminescent impurity doping, using photonic crystals to tailor the excitation field [153], architecture of a core-shell structure to suppress surface-related quenching mechanisms, utilization of metallic structures for surface plasmon enhanced upconversion [154], *etc.*; (ii) The low and narrow absorption of rare-earth ions results in harvesting of only a small fraction of sunlight for rare earth upconversion. Design of hierarchical nanostructures to incorporate a range of rare earth ions, without introducing deleterious cross relaxations, collectively can produce intense broad band upconversion [61,155]. Alternatively, external sensitizers that have strong and broad absorption, such as organic dyes [156], quantum dots [157], and transition metal ions [158] can be utilized to sensitize rare earth ions to entail upconversion; (iii) Photon upconversion is a nonlinear optical process, defining the strong dependence of upconversion luminescence on light irradiance. Most photon upconversion materials thus have limited luminescent upconverting efficiency in the range of sun irradiance (~1000 $W/m^2$). Triplet-triplet annihilation upconversion (TTA UC) has impressive efficacy under sun irradiation, favorable for uses in solar cells but with wide band-gap [101,159]. This is because the involved absorption is typically less than 700 nm [27,29,101,159,160,161,162]. Development of TTA UC with wavelengths in the IR range is appealing [102]; (iv) The upconversion in quantum nanostructures has been demonstrated at a visible wavelength, but not shown in the broad spectral range, in particular, the IR range under sunlight irradiance. This limits its uses in PV devices. Future works to overcome this problem are attractive. In all, development of upconversion materials with broad-band, strong, and tunable absorption as well as high upconverting efficiency is required, which will significantly boost solar cell efficiency by upconverison of transmitted sub-band-gap photons.

## ACKNOWLEDGMENTS

This work was supported in part by the National Science Foundation of China (No. 51102066), the National Science Fund for Distinguished Young Scholars (No. 51325201), the International Cooperation Project in the Ministry of Science and Technology (No. 2014DFA50740), the Program for Basic Research Excellent Talents in Harbin Institute of Technology, China (BRETIII 2012018), and the Fundamental Research Funds for the Central Universities, China (AUGA 5710052614).

## CONFLICTS OF INTEREST

The authors declare no conflict of interest.

## REFERENCES

1. Kalyanasundaram, K.; Gratzel, M. Themed issue: Nanomaterials for energy conversion and storage. *J. Mater. Chem.***2012**, *22*, 24190–24194.

2. Nozik, J.; Miller, J. Introduction to solar photon conversion. *Chem. Rev.* **2010**, *110*, 6443–6445.

3. Luna-Rubio, R.; Trejo-Peria, M.; Vargas-Vázquez, D.; Ríos-Moreno, G.J. Optimal sizing of renewable hybrids energy systems: A review of methodologies. *Sol. Energy* **2012**, *86*, 1077–1088.

4. Muller-Furstenberger, G.; Wagner, M. Exploring the environmental Kuznets hypothesis: Theoretical and environmental problem. *Ecol. Econ.* **2007**, *62*, 648–660.

5. Simmons, M.R. Twilight in the Desert. In *The Coming Saudi Oil Shock and the World Economy*; John Wiley & Sons: Hoboken, NJ, USA, 2005.

6. Olah, G.O.; Goepert, A.; Prakash, G.K.S. *Beyond Oil and Gas: The Methanol Economy*; Wiley-VCH: Weinheim, Germany, 2006.

7. Scholes, G.D.; Fleming, G.R.; Olaya-Castro, A.; van Grondelle, R. Lessons from nature about solar light harvesting.*Nat. Chem.* **2011**, *3*, 763–774.

8. Lewis, N.S. Toward cost-effective solar energy use. *Science* **2007**, *315*, 798–801.

9. Águas, H.; Ram, S.K.; Araújo, A.; Gaspar, D.; Vicente, A.; Filonovich, S.A.; Fortunato, E.; Martins, R.; Ferreira, I. Silicon thin film solar cells on commercial tiles. *Energy Environ. Sci.* **2011**, *4*, 4620–4632.

10. Grätzl, M. Photovoltaic and photoelectrochemical conversion of solar energy. *Philos. Trans. R. Soc. A* **2007**, *365*, 993–1005.

11. Lewis, N.S.; Nocera, D.G. Powering the planet: Chemical challenges in solar energy utilization. *Proc. Natl. Acad. Sci. USA* **2006**, *103*, 15729–15735.

12. Morton, O. Solar energy: A new day dawning? Silicon Valley Sunrise. *Nature* **2006**, *443*, 19–22.

13. Van der Ende, B.M.; Aarts, L.; Andries, M. Lanthanide ions as spectral converters for solar cells. *Phys. Chem. Chem. Phys.* **2009**, *11*, 11081–11095.

14. Saga, T. Advances in crystalline silicon solar cell technology for industrial mass production. *NPG Asia Mater.* **2010**, *2*, 96–102.

15. Chen, G.Y.; Seo, J.; Yang, C.H.; Prasad, P.N. Nanochemistry and nanomaterials for photovoltaics. *Chem. Soc. Rev.* **2013**,*42*, 8304–8338.

16. Schockley, W.; Queisser, H.J. Detailed Balance Limit of Efficiency of p-n Junction Solar Cells. *J. Appl. Phys.* **1961**, *32*, 510–519.

17. Trupke, T.; Shalav, A.; Richards, B.S.; Würfel, P.; Green, M.A. Efficiency enhancement of solar cells by luminescent up-conversion of sunlight. *Sol. Energy Mater. Sol. Cells* **2006**, *90*, 3327–3338.

18. Zhou, J.; Liu, Q.; Feng, W.; Sun, Y.; Li, F. Upconversion Luminescent Materials: Advances and Applications. *Chem. Rev.* **2015**, *115*, 395–465.

19. Haase, M.; Schäfer, H. Upconverting nanoparticles. *Angew. Chem. Int. Ed.* **2011**, *50*, 5808–5829.

20. Sun, L.D.; Wang, Y.F.; Yan, C.H. Paradigms and challenges for bioapplication of rare earth upconversion luminescent nanoparticles: Small size and tunable emission/excitation spectra. *Acc. Chem. Res.* **2014**, *47*, 1001–1009.

21. Zhou, J.; Liu, Z.; Li, F.Y. Upconversion nanophosphors for small-animal imaging. *Chem. Soc. Rev.* **2012**, *41*, 1323–1349.

22. Gorris, H.H.; Wolfbeis, O.S. Photon-upconverting nanoparticles for optical encoding and multiplexing of cells, biomolecules, and microspheres. *Angew. Chem. Int. Ed.* **2013**, *52*, 3584–3600.

23. Liu, Q.; Feng, W.; Li, F. Water-soluble lanthanide upconversion nanophosphors: Synthesis and bioimaging applications *in vivo*. *Chem. Rev.* **2014**, *273*, 100–110.

24. Feng, W.; Han, C.M.; Li, F.Y. Upconversion-Nanophosphor-Based Functional Nanocomposites. *Adv. Mater.* **2013**, *25*, 5287–5303.

25. Wang, F.; Banerjee, D.; Liu, Y.S.; Chen, X.Y.; Liu, X.G. Upconversion Nanoparticles in Biological Labeling, Imaging and Therapy. *Analyst* **2010**, *135*, 1839–1854.

26. Wang, F.; Liu, X.G. Recent advances in the chemistry of lanthanide-doped upconversion nanocrystals. *Chem. Soc. Rev.* **2009**, *38*, 976–989.

27. Zhao, J.Z.; Ji, S.M.; Guo, H.M. Triplet-triplet annihilation based upconversion: From triplet sensitizers and triplet acceptors to upconversion quantum yields. *RSC Adv.* **2011**, *1*, 937–950.

28. Soo, H.L.; Mathieu, A.A.; Roberto, V.; Christoph, W.; Yoan, C.S. Light upconversion by triplet–triplet annihilation in diphenylanthracene-based copolymers. *Polym. Chem.* **2014**, *5*, 6898–6904.

29. Singh-Rachford, T.N.; Castellano, F.N. Photon upconversion based on sensitized triplet-triplet annihilation. *Coord. Chem. Rev.* **2010**, *254*, 2560–2573.

30. Deutsch, Z.; Neeman, L.; Oron, D. Luminescence upconversion in colloidal double quantum dots. *Nat. Nanotechnol.***2013**, *8*, 649–653.

31. Deutsch, Z.; Schwartz, O.; Tenne, R.; Popovitz-Biro, R.; Oron, D. Two-color antibunching from band-gap engineered colloidal semiconductor nanocrystals. *Nano Lett.* **2012**, *12*, 2948–2952.

32. Xing, G.; Liao, Y.; Wu, X.; Chakrabortty, S.; Liu, X.; Yeow, E.K.L.; Chan, Y.; Sum, T.C. Ultralow-threshold two-photon pumped amplified spontaneous emission and lasing from seeded CdSe/CdS nanorod heterostructures. *ACS Nano***2012**, *6*, 10835–10844.

33. Bunzli, J.C.G. Benefiting from the Unique Properties of Lanthanide Ions. *Acc. Chem. Res.* **2006**, *39*, 53–61.

34. Bunzli, J.C.G. Lanthanide luminescence for biomedical analyses and imaging. *Chem. Rev.* **2010**, *110*, 2729–2755.

35. Carlos, L.D.; Ferreira, R.A.S.; Bermudez, V.D.; Julian-Lopez, B.; Escribano, P. Progress on lanthanide-based organic-inorganic hybrid phosphors. *Chem. Soc. Rev.* **2011**, *40*, 536–549.

36. Kar, A.; Patra, A. Impacts of core-shell structures on properties of lanthanide-based nanocrystals: Crystal phase, lattice strain, downconversion, upconversion and energy transfe. *Nanoscale* **2012**, *4*, 3608–3619.

37. Goesmann, H.; Feldmann, C. Nanoparticulate Functional Materials. *Angew. Chem. Int. Ed.* **2010**, *49*, 1362–1395.

38. Ohwaki, J.; Wang, Y. Efficient 1.5mm to visible upconversion in $Er^{3+}$-doped halide phosphors. *Jpn. J. Appl. Phys.* **1994**,*33*, 334–337.

39. Auzel, F.; Pecile, D.; Morin, D. Rare earth doped vitroceramics: New, efficient, blue and green emitting materials for infrared upconversion. *J. Electrochem. Soc.* **1975**, *122*.

40. Menyuk, N.; Dwight, K.; Pierce, J.W. $NaYF_4$:Yb,Er—An efficient upconversion phosphor. *Appl. Phys. Lett.* **1972**, *21*, 159–161.

41. Suyver, J.F.; Grimm, J.; Krämer, K.W.; Güdel, H.U. Highly Efficient Near-Infrared to Visible Up-Conversion Process in $NaYF_4$:$Er^{3+}$,$Yb^{3+}$. *J. Lumin.* **2005**, *114*, 53–59.

42. Krämer, K.W.; Biner, D.; Frei, G.; Güdel, H.U.; Hehlen, M.P.; Lüthi, S.T. Hexagonalsodium yttrium fluoride based green and blue emitting upconversion phosphors. *Chem. Mater.* **2004**, *16*, 1244–1251.

43. Wang, L.; Li, X.; Li, Z.; Chu, W.; Li, R.; Lin, K.; Qian, H.; Wang, Y.; Wu, C.; Li, J.; *et al.* A New Cubic Phase for a $NaYF_4$Host Matrix Offering High Upconversion Luminescence Efficiency. *Adv. Mater.* **2015**, *27*.

44. Chen, G.Y.; Ohulchanskyy, T.Y.; Kachynski, A.; Agren, H.; Prasad, P.N. Intense Visible and Near-Infrared Upconversion Photoluminescence in Colloidal $LiYF_4:Er^{3+}$ Nanocrystals under Excitation at 1490 nm. *ACS Nano* **2011**,*5*, 4981–4986.

45. Kumar, G.A.; Pokhrel, M.; Sardar, D.K. Intense visible and near infrared upconversion in $M_2O_2S$:Er (M = Y, Gd, La) phosphor under 1550 nm excitation. *Mater. Lett.* **2012**, *68*, 395–398.

46. Zheng, K.Z.; Zhao, D.; Zhang, D.S.; Liu, N.; Qin, W.P. Ultraviolet upconversion fluorescence of $Er^{3+}$ induced by 1560 nm laser excitation. *Opt. Lett.* **2010**, *35*, 2442–2444.

47. Fischer, S.; Goldschmidt, J.C.; Loper, P.; Bauer, G.H.; Bruggemann, R.; Kramer, K.; Biner, D.; Hermle, M.; Glunz, S.W. Enhancement of silicon solar cell efficiency by upconversion: Optical and electrical characterization. *J. Appl. Phys.***2010**, *108*, 044912.

48. Li, C.; Lin, J. Rare earth fluoride nano-/microcrystals: Synthesis, surface modification and application. *J. Mater. Chem.***2010**, *20*, 6831–6847.

49. Chen, G.Y.; Zhang, Y.G.; Somesfalean, G.; Zhang, Z.G.; Sun, Q.; Wang, F.P. Two-color upconversion in rare-earth-ion-doped $ZrO_2$ nanocrystals. *Appl. Phys. Lett.* **2006**, *89*, 163105.

50. Chen, G.Y.; Somesfalean, G.; Liu, Y.; Zhang, Z.G.; Sun, Q.; Wang, F.P. Upconversion mechanism for two-color emission in rare-earth-ion-doped $ZrO_2$ nanocrystals. *Phys. Rev. B* **2007**, *75*, 195204.

51. Boyer, J.C.; Johnson, N.J.J.; van Veggel, F.C.J.M. Upconverting Lanthanide-Doped $NaYF_4$-PMMA Polymer Composites Prepared by in Situ Polymerization. *Chem. Mater.* **2009**, *21*, 2010–2012.

52. Boyer, J.C.; Cuccia, L.A.; Capobianco, J.A. Synthesis of Colloidal Upconverting $NaYF_4:Er^{3+}/Yb^{3+}$ and $Tm^{3+}/Yb^{3+}$Monodisperse Nanocrystals. *Nano Lett.* **2007**, *7*, 847–852.

53. Heer, S.; Kompe, K.; Gudel, H.U.; Haase, M. Highly Efficient Multicolour Upconversion Emission in Transparent Colloids of Lanthanide-Doped $NaYF_4$ Nanocrystals. *Adv. Mater.* **2004**, *16*, 2102–2105.

54. Chen, D.; Zhou, Y.; Wan, Z.; Huang, P.; Yu, H.; Lu, H.; Ji, Z. Enhanced upconversion luminescence in phase-separation-controlled crystallization glass ceramics containing Yb/Er(Tm):$NaLuF_4$ nanocrystals. *J. Eur. Ceram. Soc.***2015**, *35*, 2129–2137.

55. Cheng, E.; Yin, W.; Bai, S.; Qiao, R.; Zhong, Y.; Li, Z. Synthesis of vis/NIR-driven hybrid photocatalysts by electrostatic assembly of $NaYF_4$:Yb, Tm nanocrystals on g-$C_3N_4$ nanosheets. *Mater. Lett.* **2015**, *146*, 87–90.

56. Chen, G.Y.; Shen, J.; Ohulchanskyy, T.Y.; Patel, N.J.; Kutikov, A.; Li, Z.P.; Song, J.; Pandey, R.K.; Agren, H.; Prasad, P.N.;*et al.* ($\alpha$-$NaYbF_4:Tm^{3+}$)/$CaF_2$ Core/Shell Nanoparticles with Efficient Near-Infrared to Near-Infrared Upconversion for High-Contrast Deep Tissue Bioimaging. *ACS Nano* **2012**, *6*, 8280–8287.

57. Li, Q.; Lin, J.; Wu, J.; Lan, Z.; Wang, Y.; Peng, F.; Huang, M. Enhancing photovoltaic performance of dye-sensitized solar cell by rare-earth doped oxide of $Lu_2O_3:(Tm^{3+},Yb^{3+})$. *Electrochim. Acta* **2011**, *56*, 4980–4984.

58. Chen, G.Y.; Ohulchanskyy, T.Y.; Kumar, R.; Agren, H.; Prasad, P.N. Ultrasmall Monodisperse $NaYF_4:Yb^{3+}/Tm^{3+}$Nanocrystals with Enhanced Near-Infrared to Near-Infrared Upconversion Photoluminescence. *ACS Nano* **2010**, *4*, 3163–3168.

59. Ramakrishna, P.V.; Pammi, S.V.N.; Samatha, K. UV-visible upconversion studies of $Nd^{3+}$ ions in lead tellurite glass.*Solid State Commun.* **2013**, *155*, 21–24.

60. Shang, Y.; Hao, S.; Liu, J.; Tan, M.; Wang, N.; Yang, C.; Chen, G. Synthesis of Upconversion $\beta$-$NaYF_4:Nd^{3+}/Yb^{3+}/Er^{3+}$Particles with Enhanced Luminescent Intensity through Control of Morphology and Phase. *Nanomaterials* **2015**, *5*, 218–232.

61. Wang, Y.-F.; Liu, G.-Y.; Sun, L.-D.; Xiao, J.-W.; Zhou, J.-C.; Yan, C.-H. $Nd^{3+}$-Sensitized Upconversion Nanophosphors: Efficient *in Vivo* Bioimaging Probes with Minimized Heating Effect. *ACS Nano* **2013**, *7*, 7200–7206.

62. Li, X.; Wang, R.; Zhang, F.; Zhou, L.; Shen, D.; Yao, C.; Zhao, D. $Nd^{3+}$ Sensitized Up/Down Converting Dual-Mode Nanomaterials for Efficient *in-vitro* and *in-vivo* Bioimaging Excited at 800 nm. *Sci. Rep.* **2013**, *3*.

63. Shalav, A.; Richards, B.S.; Trupke, T. Application of $NaYF_4:Er^{3+}$ up-converting phosphors for enhanced near-infrared silicon solar cell response. *Appl. Phys. Lett.* **2005**, *86*, 013505.

64. Tian, G.; Zheng, X.; Zhang, X.; Yin, W.; Yu, J.; Wang, D.; Zhang, Z.; Yang, X.; Gu, Z.; Zhao, Y. TPGS-stabilized $NaYbF_4$:Er upconversion nanoparticles for dual-modal fluorescent/CT imaging and anticancer drug delivery to overcome multi-drug resistance. *Biomaterials* **2015**, *40*, 107–116.

65. Chen, G.Y.; Qiu, H.L.; Fan, R.W.; Hao, S.W.; Tan, S.; Yang, C.H.; Han, G. Lanthanide-doped ultrasmall yttrium fluoride nanoparticles with enhanced multicolor upconversion photoluminescence. *J. Mater. Chem.* **2012**, *22*, 20190.

66. Jung, T.; Jo, H.L.; Nam, S.H.; Yoo, B.; Cho, Y.; Kim, J.; Kim, H.M.; Hyeon, T.; Suh, Y.D.; Lee, H.; *et al.* The preferred upconversion pathway for the red emission of lanthanide-doped upconverting nanoparticles, $NaYF_4:Yb^{3+},Er^{3+}$. *Phys. Chem. Chem. Phys.* **2015**, *17*, 13201.

67. Tian, D.; Gao, D.; Chongb, B.; Liu, X. Upconversion improvement by the reduction of $Na^+$-vacancies in $Mn^{2+}$ doped hexagonal $NaYbF_4:Er^{3+}$ nanoparticles. *Dalton Trans.* **2015**, *44*, 4133.

68. Khan, A.F.; Yadav, R.; Mukhopadhya, P.K.; Singh, S.; Dwivedi, C.; Dutta, V.; Chawla, S. Core-shell nanophosphor with enhanced NIR–visible upconversion as spectrum modifier for enhancement of solar cell efficiency. *J. Nanopart. Res.*
**2011**, *13*, 6837–6846.

69. Chen, Y.; Liu, B.; Deng, X.; Huang, S.; Hou, Z.; Li, C.; Lin, J. Multifunctional $Nd^{3+}$-sensitized upconversion nanomaterials for synchronous tumor diagnosis and treatment. *Nanoscale* **2015**, *7*, 8574.

70. Zhang, S.; Wang, J.; Xu, W.; Chen, B.; Yu, W.; Xu, L.; Song, H. Fluorescence Resonance energy transfer between $NaYF_4$:Yb,Tm upconversion nanoparticles and gold nanorods: Near-infrared responsive biosensor for streptavidin. *J. Lumin.* **2014**, *147*, 278–283.

71. Luu, Q.; Hor, A.; Fisher, J.; Anderson, R.B.; Liu, S.; Luk, T.; Paudel, H.P.; Baroughi, M.F.; May, P.S.; Smith, S. Two-Color Surface Plasmon Polariton Enhanced Upconversion in $NaYF_4$:Yb:Tm Nanoparticles on Au Nanopillar Arrays. *J. Phys. Chem. C* **2014**, *118*, 3251–3257.

72. Lin, H.; Xu, D.; Teng, D.; Yang, S.; Zhang, Y. Simultaneous size and luminescence control of $NaYF_4$:$Yb^{3+}$/$RE^{3+}$ (RE = Tm, Ho) microcrystals via $Li^+$ doping. *Opt. Mater.* **2015**, *45*, 229–234.

73. Amemori, S.; Yanai, N.; Kimizuka, N. Metallonaphthalocyanines as triplet sensitizers for near-infrared photon upconversion beyond 850 nm. *Phys. Chem. Chem. Phys.* **2015**, *17*, 22557–22560.

74. Huang, Z.; Li, X.; Mahboub, M.; Hanson, K.M.; Nichols, V.M.; Le, H.; Tang, M.L.; Bardeen, C.J. Hybrid Molecule—Nanocrystal Photon Upconversion across the Visible and Near-Infrared. *Nano Lett.* **2015**, *15*, 5552–5557.

75. Monguzzi, A.; Braga, D.; Gandini, M.; Holmberg, V.C.; Kim, D.K.; Sahu, A.; Norris, D.J.; Meinardi, A.F. Broadband Up-Conversion at Subsolar Irradiance: Triplet-Triplet Annihilation Boosted by Fluorescent Semiconductor Nanocrystals. *Nano Lett.* **2014**, *14*, 6644–6650.

76. Monguzzi, A.; Borisov, S.; Pedrini, J.; Klimant, I.; Salvalaggio, M.; Biagini, P.; Melchiorre, F.; Lelii, C.; Meinardi, F. Efficient Broadband Triplet-Triplet Annihilation-Assisted Photon Upconversion at Subsolar Irradiance in Fully Organic Systems. *Adv. Funct. Mater.* **2015**, *25*.

77. Mahato, P.; Monguzzi, A.; Yanai1, N.; Yamada, T.; Kimizuka, N. Fast and long-range triplet exciton diffusion in metal-organic frameworks for photon upconversion at ultralow excitation power. *Nat. Mater.* **2015**, *14*, 924–930.

78. Duan, P.; Yanai, N.; Kimizuka, N. Photon Upconverting Liquids: Matrix-Free Molecular Upconversion Systems Functioning in Air. *J. Am. Chem. Soc.* **2013**, *135*, 19056–19059.

79. Duan, P.; Yanai, N.; Nagatomi, H.; Kimizuka, N. Photon Upconversion in Supramolecular Gel Matrixes: Spontaneous Accumulation of Light-Harvesting Donor—Acceptor Arrays in Nanofibers and Acquired Air Stability. *J. Am. Chem. Soc.* **2015**, *137*, 1887–1894.

80. Ogawa1, T.; Yanai, N.; Monguzzi, A.; Kimizuka1, N. Highly Efficient Photon Upconversion in Self-Assembled Light-Harvesting Molecular Systems. *Sci. Rep.* **2015**, *5*.

81. Hisamitsu, S.; Yanai, N.; Kimizuka, N. Photon-Upconverting Ionic Liquids: Effective Triplet Energy Migration in Contiguous Ionic Chromophore Arrays. *Angew. Chem. Int. Ed.* **2015**, *54*, 1–6.

82. Vadrucci, R.; Weder, C.; Simon, Y.C. Low-power photon upconversion in organic glasses. *J. Mater. Chem. C* **2014**, *2*, 2837–2841.

83. Vadrucci, R.; Weder, C.; Simon, Y.C. Organogels for low-power light upconversion. *Mater. Horiz.* **2015**, *2*, 120–124.

84. Green, M.A.; Emery, K.; Hishikawa, Y.; Warta, W.; Dunlop, E.D. Solar cell efficiency tables. *Prog. Photovolt.* **2014**, *22*, 701–710.

85. Lahoz, F. $Ho^{3+}$-doped nanophase glass ceramics for efficiency enhancement in silicon solar cells. *Opt. Lett.* **2008**, *33*, 2982–2984.

86. Lahoz, F.; Perez-Rodriguez, C.; Hernandez, S.E.; Martin, I.R.; Lavin, V.; Rodriguez-Mendoza, U.R. Upconversion mechanisms in rare-earth doped glasses to improve the efficiency of silicon solar cells. *Sol. Energy Mater. Sol. Cells***2011**, *95*, 1671–1677.

87. Chen, D.; Lei, L.; Yang, A.; Wang, Z.; Wang, Y. Ultra-broadband near-infrared excitable upconversion core/shell nanocrystals. *Chem. Commun.* **2012**, *48*, 5898–5900.

88. Shao, W.; Chen, G.; Ohulchanskyy, T.Y.; Kuzmin, A.; Damasco, J.; Qiu, H.; Yang, C.; Hans, Å.; Prasad, P.N. Lanthanide-Doped Fluoride Core/Multishell Nanoparticles for Broad-Band Upconversion of Infrared Light. *Adv. Opt. Mater.* **2015**,*3*, 575–582.

89. Dobrovolsky, A.; Sukrittanon, S.; Kuang, Y.; Tu, C.W.; Chen, W.M.; Buyanova, I.A. Energy Upconversion in GaP/GaNP Core/Shell Nanowires for Enhanced Near-Infrared Light Harvesting. *Small* **2014**, *10*, 4403–4406.

90. Zhang, X.D.; Jin, X.; Wang, D.F.; Xiong, S.Z.; Geng, X.H.; Zhao, Y. Synthesis of $NaYF_4$:Yb, Er nanocrystals and its application in silicon thin film solar cells. *Phys. Status Solidi C* **2010**, *7*, 1128–1131.

91. De Wild, J.; Meijerink, A.; Rath, J.K.; van Sark, W.G.J.H.M.; Schropp, R.E.I. Towards upconversion for amorphous silicon solar cells. *Sol. Energy Mater. Sol. Cells* **2010**, *94*, 1919–1922.

92. De Wild, J.; Rath, J.K.; Meijerink, A.; van Sark, W.G.J.H.M.; Schropp, R.E.I. Enhanced near-infrared response of a-Si:H solar cells with β-$NaYF_4$:$Yb^{3+}$(18%), $Er^{3+}$(2%) upconversion phosphors. *Sol. Energy Mater. Sol. Cells* **2010**, *94*, 2395–2398.

93. Yang, F.; Liu, C.; Wei, D.; Chen, Y.; Lu, J.; Yang, S. $Er^{3+}$–$Yb^{3+}$ co-doped $TeO_2$-$PbF_2$ oxyhalide tellurite glasses for amorphous silicon solar cells. *Opt. Mater.* **2014**, *36*, 1040–1043.

94. Kumar, P.; Gupta, B.K. New insight into rare-earth doped gadolinium molybdate nanophosphor assisted broad spectral converters from UV to NIR for silicon solar cells. *RSC Adv.* **2015**, *5*, 24729–24736.

95. Kik, P.; Polman, A. Cooperative upconversion as the gain-limiting factor in Er doped miniature $Al_2O_3$ optical waveguide amplifiers. *J. Appl. Phys.* **2003**, *93*, 5008–5012.

96. Pollack, S.; Chang, D. Ion-pair upconversion pumped laser emission in $Er^{3+}$ ions in YAG, YLF, $SrF_2$, and $CaF_2$ crystals.*J. Appl. Phys.* **1988**, *64*, 2885–2893.

97. Wang, X.F.; Yan, X.H.; Kan, C.X. Controlled synthesis and optical characterization of multifunctional ordered $Y_2O_3$:$Er^{3+}$ porous pyramid arrays. *J. Mater. Chem.* **2011**, *21*, 4251–4256.

98. Ivanova, S.; Pelle, F. Strong 1.53 μm to NIR-VIS-UV upconversion in Er-doped fluoride glass for high-efficiency solar cells. *J. Opt. Soc. Am. B* **2009**, *26*, 1930–1938.

99. Ohwaki, J.; Wang, Y. New efficient upconversion phosphor $BaCl_2$:Er under 1.5 mm excitation. *Electron. Lett.* **1993**, *29*, 351–352.

100. Chen, Y.; He, W.; Jiao, Y.; Wang, H.; Hao, X.; Lu, J.; Yang, S.E. β-NaYF4:$Er^{3+}$ (10%) microprisms for the enhancement of a-Si:H solar cell near-infrared responses. *J. Lumin.* **2012**, *132*, 2247–2250.

101. Cheng, Y.Y.; Fückel, B.; MacQueen, R.W.; Khoury, T.; Clady, R.G.; Schulze, T.F.; Ekins-Daukes, N.J.; Crossley, M.J.; Stannowski, B.; Lips, K.; *et al.* Improving the light-harvesting of amorphous silicon solar cells with photochemical upconversion. *Energy Environ. Sci.* **2012**, *5*, 6953–6959.

102. Schulze, T.F.; Schmidt, T.W. Photochemical upconversion: Present status and prospects for its application to solar energy conversion. *Energy Environ. Sci.* **2015**, *8*, 103–125.

103. Nattestad, A.; Cheng, Y.Y.; MacQueen, R.W.; Schulze, T.F.; Thompson, F.W.; Mozer, A.J.; Fückel, B.; Khoury, T.; Crossley, M.J.; Lips, K.; *et al.* Dye-Sensitized Solar Cell with Integrated Triplet–Triplet Annihilation Upconversion System. *J. Phys. Chem. Lett.* **2013**, *4*, 2073–2078.

104. Schulze, T.F.; Cheng, Y.Y.; Fückel, B.; MacQueen, R.W.; Danos, A.; Davis, N.J.L.K.; Tayebjee, M.J.Y.; Khoury, T.; Clady, R.G.C.R.; Ekins-Daukes, N.J.; *et al.* Photochemical Upconversion Enhanced Solar Cells: Effect of a Back Reflector. *Aust. J. Chem.* **2012**, *65*, 480–485.

105. Schulze, T.F.; Czolk, J.; Cheng, Y.; Fuckel, B.; MacQueen, R.W.; Khoury, T.; Crossley, M.J.; Stannowski, B.; Lips, K.; Lemmer, U.; *et al.* Efficiency Enhancement of Organic and Thin-Film Silicon Solar Cells with Photochemical Upconversion. *J. Phys. Chem. C* **2012**, *116*, 22794–22801.

106. Schulze, T.F.; Cheng, Y.Y.; Khoury, T.; Crossley, M.J.; Stannowski, B.; Lips, K.; Schmidt, T.W. Micro-optical design of photochemical upconverters for thin-film solar cells. *J. Photonics Energy* **2013**, *3*, 034598.

107. George, B.M.; Behrends, J.; Schnegg, A.; Schulze, T.F.; Fehr, M.; Korte, L.; Rech, B.; Lips, K.; Rohrmüller, M.; Rauls, E.;*et al.* Atomic Structure of Interface States in Silicon Heterojunction Solar Cells. *Phys. Rev. Lett.* **2013**, *110*, 136803.

108. Gibart, P.; Auzel, F.; Guillaume, J.-C.; Zahraman, K. Below band gap IR response of substrate-free GaAs solar cells using two-photon upconversion. *Jpn. J. Appl. Phys.* **1996**, *351*, 4401–4402.

109. Lin, H.Y.; Chen, H.N.; Wu, T.H.; Wu, C.S.; Su, Y.K.; Chu, S.Y. Investigation of Green Up-Conversion Behavior in $Y_6W_2O_{15}:Yb^{3+},Er^{3+}$

Phosphor and Its Verification in 973 nm Laser-Driven GaAs Solar Cell. *J. Am. Ceram. Soc.* **2012**, *95*, 3172–3179.

110. O'Regan, B.; Grätzel, M. A low-cost, high-efficiency solar cell based on dye-sensitized. *Nature* **1991**, *353*, 737–740.

111. Grätzel, M. Photoelectrochemical cells. *Nature* **2001**, *414*, 338–344.

112. Bisquert, J. Dilemmas of Dye-Sensitized Solar Cell. *Chem. Phys. Chem.* **2011**, *12*, 1633–1636.

113. Basham, J.I.; Mor, G.K.; Grimes, C.A. Förster Resonance Energy Transfer in Dye-Sensitized Solar Cells. *ACS Nano***2010**, *4*, 1253–1258.

114. Chen, C.-Y.; Wang, M.; Li, J.-Y.; Pootrakulchote, N.; Alibabaei, L.; Ngoc-le, C.-H.; Decoppet, J.-D.; Tsai, J.-H.; Graetzel, C.; Wu, C.G.; *et al.* Highly Efficient Light-Harvesting Ruthenium Sensitizer for Thin-Film Dye-Sensitized Solar Cells.*ACS Nano* **2009**, *3*, 3103–3109.

115. Huang, X.Y.; Wang, J.X.; Yu, D.C.; Ye, S.; Zhang, Q.Y.; Sun, X.W. Spectral conversion for solar cell efficiency enhancement using $YVO_4:Bi^{3+}$, $Ln^{3+}$ (Ln = Dy, Er, Ho, Eu, Sm, and Yb) phosphors. *J. Appl. Phys.* **2011**, *109*, 113526.

116. Bünzli, J.C.G.; Eliseeva, S.V. Lanthanide NIR luminescence for telecommunications, bioanalyses and solar energy conversion. *J. Rare Earths* **2010**, *28*, 824–842.

117. Hafez, H.; Saif, M.; Abdel-Mottaleb, M.S.A. Down-converting lanthanide doped $TiO_2$ photoelectrodes for efficiency enhancement of dye-sensitized solar cells. *J. Power Sources* **2011**, *196*, 5792–5796.

118. Levinson, R.; Berdahl, P.; Akbari, H. Solar spectral optical properties of pigments—Part I: Model for deriving scattering and absorption coefficients from transmittance and reflectance measurements. *Sol. Energy Mater. Sol. C* **2005**,*89*, 319–349.

119. Hara, K.; Sato, T.; Katoh, R.; Furube, A.; Ohga, Y.; Shinpo, A.; Suga, S.; Sayama, K.; Sugihara, H.; Arakawa, H. Molecular Design of Coumarin Dyes for Efficient Dye-Sensitized Solar Cells. *J. Phys. Chem. B* **2003**, *107*, 597–606.

120. Campbell, W.M.; Burrell, A.K.; Officer, D.L.; Jolley, K.W. Porphyrins as light harvesters in the dye-sensitised $TiO_2$ solar cell. *Coord. Chem. Rev.* **2004**, *248*, 1363–1379.

121. Shan, G.-B.; Demopoulos, G.P. Near-Infrared Sunlight Harvesting in Dye-Sensitized Solar Cells via the Insertion of an Upconverter-$TiO_2$ Nanocomposite Layer. *Adv. Mater.* **2010**, *22*, 4373–4377.

122. Shan, G.B.; Assaaoudi, H.; Demopoulos, G.P. Enhanced Performance of Dye-Sensitized Solar Cells by Utilization of an External, Bifunctional Layer Consisting of Uniform β-$NaYF_4$:$Er^{3+}$/$Yb^{3+}$ Nanoplatelets. *ACS Appl. Mater. Interfaces* **2011**,*3*, 3239–3243.

123. Yuan, C.Z.; Chen, G.Y.; Prasad, P.N.; Ohulchanskyy, T.Y.; Ning, Z.J.; Tian, H.; Sund, L.C.; Agren, H. Use of colloidal upconversion nanocrystals for energy relay solar cell light harvesting in the near-infrared region. *J. Mater. Chem.* **2012**,*22*, 16709–16713.

124. Li, L.; Yang, Y.; Fan, R.; Jiang, Y.; Wei, L.; Shi, Y.; Yu, J.; Chen, S.; Wang, P.; Yang, B.; *et al.* A simple modification of near-infrared photon-to-electron response with fluorescence resonance energy transfer for dye-sensitized solar cells. *J. Power Sources* **2014**, *264*, 254–261.

125. Wang, J.; Lin, J.; Wu, J.; Huang, M.; Lan, Z.; Chen, Y.; Tang, S.; Fan, L.; Huang, Y. Application of $Yb^{3+}$, $Er^{3+}$-doped yttrium oxyfluoride nanocrystals in dye-sensitized solar cells. *Electrochim. Acta* **2012**, *70*, 131–135.

126. Zhang, J.; Shen, H.; Guo, W.; Wang, S.; Zhu, C.; Xue, F.; Hou, J.; Su, H.; Yuan, Z. An upconversion $NaYF_4$:$Yb^{3+}$,$Er^{3+}$/$TiO_2$ coreeshell nanoparticle photoelectrode for improved efficiencies of dye-sensitized solar cells. *J. Power Sources* **2013**, *226*, 47–53.

127. Liang, L.L.; Liu, Y.M.; Zhao, X.Z. Double-shell beta-NaYF4:$Yb^{3+}$,$Er^{3+}$/$SiO_2$/$TiO_2$ submicroplates as a scattering and upconverting layer for efficient dye-sensitized solar cells. *Chem. Commun.* **2013**, *49*, 3958–3960.

128. Dyck, N.C.; Demopoulos, G.P. Integration of upconverting β-$NaYF_4$:$Yb^{3+}$,$Er^{3+}$@$TiO_2$ composites as light harvesting layers in dye-sensitized solar cells. *RSC Adv.* **2014**, *4*, 52694–52701.

129. Wu, J.H.; Wang, J.L.; Lin, J.M.; Lan, Z.; Tang, Q.W.; Huang, M.L.; Huang, Y.F.; Fan, L.Q.; Li, Q.B.; Tang, Z.Y. Enhancement of the Photovoltaic Performance of Dye-Sensitized Solar Cells by Doping $Y_{0.78}Yb_{0.20}Er_{0.02}F_3$ in the Photoanode. *Adv. Energy Mater.* **2012**, *2*, 78–81.

130. Liu, M.; Lu, Y.; Xie, Z.B.; Chow, G.M. Enhancing near-infrared solar cell response using upconverting transparentceramics. *Sol. Energy Mater. Sol. Cells* **2011**, *95*, 800–803.

131. Wang, J.; Wu, J.; Lin, J.; Huang, M.; Huang, Y.; Lan, Z.; Xiao, Y.; Yue, G.; Yin, S.; Sato, T. Application of $Y_2O_3:Er^{3+}$Nanorods in Dye-Sensitized Solar Cells. *ChemSusChem* **2012**, *5*, 1307–1312.

132. Bai, J.; Zhao, R.; Han, G.; Li, Z.; Diao, G. Synthesis of 1D upconversion $CeO_2$:Er,Yb nanofibers via electrospinning and their performance in dye-sensitized solar cells. *RSC Adv.* **2015**, *5*, 43328–43333.

133. Yuan, C.; Chen, G.; Li, L.; Damasco, J.A.; Ning, Z.; Xing, H.; Zhang, T.; Sun, L.; Zeng, H.; Cartwright, A.N.; *et al.* Simultaneous Multiple Wavelength Upconversion in a Core-Shell Nanoparticle for Enhanced Near Infrared Light Harvesting in a Dye-Sensitized Solar Cell. *ACS Appl. Mater. Interfaces* **2014**, *6*, 18018–18025.

134. Zhao, P.; Zhu, Y.; Yang, X.; Jiang, X.; Shen, J.; Li, C. Plasmon-enhanced efficient dye-sensitized solar cells using core–shell-structured β-$NaYF_4$:Yb,Er@$SiO_2$@Au nanocomposites. *J. Mater. Chem. A* **2014**, *2*, 16523–16530.

135. Li, L.; Yang, Y.; Fan, R.; Chen, S.; Wang, P.; Yang, B.; Cao, W. Conductive Upconversion Er, Yb-FTO Nanoparticle Coating To Replace Pt as a Low-Cost and High-Performance Counter Electrode for Dye-Sensitized Solar Cells. *ACS Appl. Mater. Interfaces* **2014**, *6*, 8223–8229.

136. Yu, J.; Yang, Y.; Fan, R.; Liu, D.; Wei, L.; Chen, S.; Li, L.; Yang, B.; Cao, W. Enhanced Near-Infrared to Visible Upconversion Nanoparticles of $Ho^{3+}$-$Yb^{3+}$-F Tri-Doped $TiO_2$ and Its Application in Dye-Sensitized Solar Cells with 37% Improvement in Power Conversion Efficiency. *Inorg. Chem.* **2014**, *53*, 8045–8053.

137. Boudreault, P.L.T.; Najari, A.; Leclerc, M. Processable Low-Bandgap Polymers for Photovoltaic Applications. *Chem. Mater.* **2011**, *23*, 456–469.

138. Chen, H.Y.; Hou, J.H.; Zhang, S.Q.; Liang, Y.Y.; Yang, G.W.; Yang, Y.; Yu, L.P.; Wu, Y.; Li, G. Polymer solar cells with enhanced open-circuit voltage and efficiency. *Nat. Photonics* **2009**, *3*, 649–653.

139. Hains, A.W.; Liang, Z.Q.; Woodhouse, M.A.; Gregg, B.A. Molecular Semiconductors in Organic Photovoltaic Cells.*Chem. Rev.* **2010**, *110*, 6689–6735.

140. Chen, J.W.; Cao, Y. Development of Novel Conjugated Donor Polymers for High-Efficiency Bulk-Heterojunction Photovoltaic Devices. *Acc. Chem. Res.* **2009**, *42*, 1709–1718.

141. He, Z.; Zhong, C.; Su, S.; Xu, M.; Wu, H.; Cao, Y. Enhanced power-conversion efficiency in polymer solar cells using an inverted device structure. *Nat. Photonics* **2012**, *6*, 591–595.

142. You, J.; Dou, L.; Yoshimura, K.; Kato, T.; Ohya, K.; Moriarty, T.; Emery, K.; Chen, C.-C.; Gao, J.; Li, G.; *et al.* A polymer tandem solar cell with 10.6% power conversion efficiency. *Nat. Commun.* **2013**, *4*, 1446.

143. Grätzel, M. Recent Advances in Sensitized Mesoscopic Solar Cells. *Acc. Chem. Res.* **2009**, *42*, 1788–1798.

144. Yella, A.; Mai, C.-L.; Zakeeruddin, S.M.; Chang, S.-N.; Hsieh, C.-H.; Yeh, C.-Y.; Grätzel, M. Molecular Engineering of Push—Pull Porphyrin Dyes for Highly Efficient Dye-Sensitized Solar Cells: The Role of Benzene Spacers. *Angew. Chem.* **2014**, *126*, 3017–3021.

145. Günes, S.; Neugebauer, H.; Sariciftci, N.S. Conjugated Polymer-Based Organic Solar Cells. *Chem. Rev.* **2007**, *107*, 1324–1338.

146. Wang, H.Q.; Batentschuk, M.; Osvet, A.; Pinna, L.; Brabec, C.J. Rare-Earth Ion Doped Up-Conversion Materials for Photovoltaic Applications. *Adv. Mater.* **2011**, *23*, 2675–2680.

147. Wang, H.Q.; Stubhan, T.; Osvet, A.; Litzov, I.; Brabec, C.J. Up-conversion semiconducting $MoO_3$:Yb/Er nanocomposites as buffer layer in organic solar cells. *Sol. Energy Mater. Sol. Cells* **2012**, *105*, 196–201.

148. Wu, J.L.; Chen, F.-C.; Chang, S.-H.; Tan, K.-S.; Tuan, H.-Y. Upconversion effects on the performance of near-infrared laser-driven polymer photovoltaic devices. *Org. Electron.* **2012**, *13*, 2104–2108.

149. Adikaari, A.A.D.; Etchart, I.; Guering, P.H.; Berard, M.; Silva, S.R.P.; Cheetham, A.K.; Curry, R.J. Near infrared up-conversion in organic photovoltaic devices using an efficient $Yb^{3+}$:$Ho^{3+}$ Co-doped $Ln_2BaZnO_5$ (Ln = Y, Gd) phosphor.*J. Appl. Phys.* **2012**, *111*.

150. Guo, W.; Zheng, K.; Xie, W.; Sun, L.; Shen, L.; Liu, C.; He, Y.; Zhang, Z. Efficiency enhancement of inverted polymer solar cells by doping $NaYF_4$:$Yb^{3+}$, $Er^{3+}$ nanocomposites in PCDTBT:PCBM active layer. *Sol. Energy Mater. Sol. Cells***2014**, *124*, 126–132.

151. Yakutkin, V.; Aleshchenkov, S.; Chernov, S.; Miteva, T.; Nelles, G.; Cheprakov, A.; Baluschev, S. Towards the IR Limit of the Triplet–Triplet Annihilation-Supported Up-Conversion: Tetraanthraporphyrin. *Chem. Eur. J.* **2008**, *14*, 9846–9850.

152. Boyer, J.C.; van veggel, F.C.J.M. Absolute quantum yield measurements of colloidal $NaYF_4$:$Er^{3+}$,$Yb^{3+}$ upconverting nanoparticles. *Nanoscale* **2010**, *2*, 1417–1419.

153. Johnson, C.M.; Reece, P.J.; Conibeer, G.J. Slow-light-enhanced upconversion for photovoltaic applications in one-dimensional photonic crystals. *Opt. Lett.* **2011**, *36*, 3990–3992.

154. Zhang, W.; Ding, F.; Chou, S.Y. Large Enhancement of Upconversion Luminescence of $NaYF_4$:$Yb^{3+}$/$Er^{3+}$ Nanocrystal by 3D Plasmonic Nano-Antennas. *Adv. Mater.* **2012**, *24*, OP236–OP241.

155. Xie, X.; Gao, N.; Deng, R.; Sun, Q.; Xu, Q.-H.; Liu, X. Mechanistic Investigation of Photon Upconversion in $Nd^{3+}$-Sensitized Core–Shell Nanoparticles. *J. Am. Chem. Soc.* **2013**, *135*, 12608–12611.

156. Zou, W.; Visser, C.; Maduro, J.A.; Pshenichnikov, M.S.; Hummelen, J.C. Broadband dye-sensitized upconversion of near-infrared light. *Nat. Photonics* **2012**, *6*, 560–564.

157. Pan, A.C.; del Canizo, C.; Canavos, E.; Santos, N.M.; Leitao, J.P.; Luque, A. Enhancement of up-conversion efficiency by combining rare earth-doped phosphors with PbS quantum dots. *Sol. Energy Mater. Sol. Cells* **2010**, *94*, 1923–1926.

158. Suyver, J.F.; Aebischer, A.; Biner, D.; Gerner, P.; Grimm, J.; Heer, S.; Kramer, K.W.; Reinhard, C.; Gudel, H.U. Novel materials doped with trivalent lanthanides and transition metal ions showing near-infrared to visible photon upconversion. *Opt. Mater.* **2005**, *27*, 1111–1130.

159. Simon, Y.C.; Weder, C. Low-Power Photon Upconversion through Triplet-Triplet Annihilation in Polymers. *J. Mater. Chem.* **2012**, *22*, 20817–20830.

160. Singh-Rachford, T.N.; Nayak, A.; Muro-Small, M.L.; Goeb, S.; Therien, M.J.; Castellano, F.N. Supermolecular-Chromophore-Sensitized Near-Infrared-to-Visible Photon Upconversion. *J. Am. Chem. Soc.* **2010**, *132*, 14203–14211.

161. Fückel, B.; Roberts, D.A.; Cheng, Y.Y.; Clady, R.; Piper, R.B.; Ekins-Daukes, N.J.; Crossley, M.J.; Schmidt, T.W. Singlet Oxygen Mediated Photochemical Upconversion of NIR Light. *J. Phys. Chem. Lett.* **2011**, *2*, 966–971.

162. Singh-Rachford, T.N.; Castellano, F.N. Nonlinear photochemistry squared: Quartic light power dependence realized in photon upconversion. *J. Phys. Chem. A* **2009**, *113*, 9266–9269.

# Chapter 6

# A cost roadmap for silicon heterojunction solar cells

**AtseLouwen [a,n], WilfriedvanSark [a], RuudSchropp [b], AndréFaaij [c]**

[a]Utrecht University, Copernicus Institute of Sustainable Development, Utrecht, The Netherlands

[b] Eindhoven University of Technology (TU/e), AppliedPhysics, Plasmaand Materials Processing, Eindhoven, TheNetherlands

c Groningen University, Energy Academy and Energy & Sustainability Research Institute,Groningen,TheNetherlands

## ABSTRACT

Research and development of silicon heterojunction (SHJ) solar cells has seen a marked increase since the recent expiry of core patents describing SHJ technology. SHJ solar cells are expected to offer various cost benefits compared to conventional crystalline silicon solar cells. This paper analyses the production costs associated with five different SHJ cell designs, including an interdigitated back-contacted (IBC) design. Using life-cycle costing, we analyzed the current cost breakdown of these SHJ designs, and compared them to conventional diffused junction monocrystalline silicon modules. Coupling the results for current designs with literature data on technological improvements, we also present a prospective analysis of production costs for the five SHJ cells and modules.

For current designs, module costs were calculated to be 0.48–0.56 USD per Watt-peak ($W_p$) for SHJ modules, compared to 0.50 USD/$W_p$ for a conventional c-Si module. The efficiency bonus for SHJ modules compared to conventional c-Si modules is offset by a strong increase in metallization costs for SHJ designs, as comparatively large amounts of low-temperature silver-paste are required. For module materials, the requirement for conductive adhesives results in a small cost penalty for SHJ modules compared to c-Si modules, which is more than balanced by the effect of higher efficiency in SHJ modules.

Our prospective study showed that improvements in cell processing and module design could result in a significant drop in production costs for all module types studied. The SHJ modules gain much advantage by reducing and replacing silver consumption, increased cell efficiency and thinner wafers and have prospective production costs of 0.29–0.35 USD/$W_p$. Conventional c-Si module cost is less sensitive to silver paste consumption, limiting the potential for cost reduction, and has prospective production costs of 0.33 USD/$W_p$. Replacement of indium-tin-oxide was not found to contribute substantially to a reduction in module costs.

## KEYWORDS

Photovoltaics; Silicon heterojunction; SHJ; Crystalline silicon; Economics; Solar; Life cycle costing

## 1. INTRODUCTION

Concurrently with the strong growth in PV module production and sales, average PV module prices have dropped sharply over the last decade. Polysilicon, wafer, cell and module prices dropped especially sharp over the last few years, as shown in Fig. 1. In the Netherlands, PV module prices including tax dropped by almost 50% between 2011 and 2013, from 2 EUR per $W_p$ to 1.13 EUR per $W_p$[1], while global spot prices (excluding tax) for PV modules have dropped to 0.6 USD/$W_p$ (see Fig. 1). Price decreases have long been following a learning curve that has been valid for multiple decades, however, more recently, due to decreased demand and resulting oversupply, prices have dropped below what could be extrapolated from the learning curve. As a result, PV producers are scrambling for opportunities to reduce production costs. On the other hand, although residential grid parity has been reached in several countries worldwide [2], PV electricity is as of yet not competitive with fossil electricity generation [3], [4], [5], [6] and [7]. These two factors emphasize the need for further cost reductions in the PV industry, in order to assure PV production that is financially sustainable and competitive with bulk electricity generation.

Photovoltaic systems offer us the possibility to produce electricity with low emissions of greenhouse gasses [9], [10], [11], [12] and [13], low energy pay-back time, and low emissions of toxic or otherwise harmful substances, compared to traditional forms of electricity production. Its' modular nature allows for the application on a variety of scales, from small-scale decentralized and off-grid to large-scale, centralized electricity production. According to International Energy Agency's World Energy Outlook, PV will contribute significantly to a sustainable energy supply system [14]. Because of this expectation, adoption of PV is being supported by national governments worldwide, at sometimes high financial costs for society [15]. Cost reductions are thus of substantial societal importance when deployment of solar energy covers larger shares of total electricity generation.

In order for PV to become a competitive source of electricity production, the levelized cost of electricity (LCOE) from PV should decrease below residential electricity prices ("socket parity") and below wholesale electricity prices ("grid parity"). Furthermore, for PV to become a viable investment option for both consumers and businesses, the payback time (PBT) of investing in PV should drop to about 3–5 years [16]. Fig. 2 shows the effect of PV system price and annual yield on LCOE and PBT. From this figure, we can deduce that, in order for PV to reach grid parity (instead of the already achieved socket parity) PV system prices still need to drop significantly. For PV to compete with combined-cycle natural gas and coal with a levelized cost of electricity (LCOE) of about 0.05 USD/kWh[17], we estimate that PV system prices need to drop below 0.60–1.00 USD/$W_p$, thus PV module prices should drop below 0.3–0.5 USD $W_p$[1]

For a large part, cost reductions in PV production have been achieved due to economies of scale, and technological learning in the PV production supply chain [18]. More recently we have seen an increased focus on intrinsic cost reductions. General approaches for cost reductions have traditionally been to decrease material use or replace expensive materials with cheaper ones. For instance, silicon consumption per watt-peak ($W_p$) has decreased significantly due to increased efficiencies and the use of increasingly thin wafers, while silver use for metallization has also decreased over the years. Significant cost reductions have been obtained with this approach, however, more recently the room for further improvement has decreased. This has resulted in a variety of approaches being researched, including a shift from the traditional diffused junction crystalline PV devices, towards alternative designs or technologies.

One of those "new" technologies is the silicon heterojunction (SHJ) solar cell technology. SHJ solar cells are produced from silicon wafers in a low temperature process that does not exceed 200 °C. High temperature diffusion of the p-n junction is replaced with a low temperature deposition of a p-doped amorphous silicon layer on an n-type monocrystalline silicon wafer.

This technology is only produced on a large scale by Panasonic (by acquiring Sanyo), but a recent expiry of the core patents describing their SHJ technology has lead to a marked increase in R&D on this technology [19] and [20]. The large interest is mainly due to the fact that [19] and [20]: (**1**) SHJ fabrication is a simple process with high efficiency cells as a result; (**2**) the deposition of the thin film layers for SHJ cells can benefit from ample experience with these processes in the flat-display industry; (**3**) SHJ modules have a low temperature coefficient which leads to higher energy yields compared to conventional c-Si modules, and (**4**) SHJ cells benefit more from the application of thinner wafers, because the deposited thin-film layers allow for very good passivation of the wafer surface.

Current R&D mainly focuses on improving device performance, but novel design structures and processing steps are also being investigated, aiming to lower production costs. Examples are alternative materials for the transparent conductive oxide (TCO), alternative metallization schemes and materials, and alternative passivation structures and materials [20]. Current SHJ cell designs rely on expensive materials (silver, indium) and processing steps.

In order to improve the economic performance of SHJ technology, the FLASH$^2$programme investigates several alternative SHJ cell designs. The research focuses not only on replacement of expensive and high price volatility materials like indium (for the TCO) and silver (for metallization) but also on simplifying the production process.

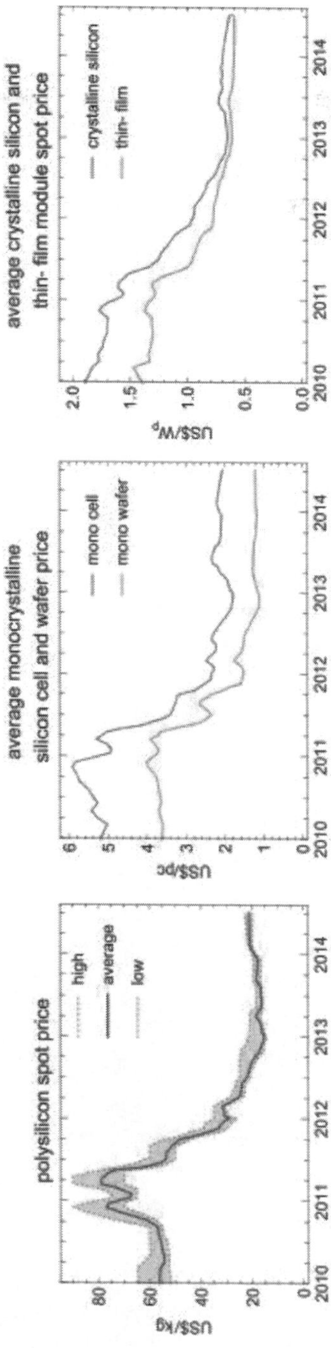


**Figure 1.** Overview of development of prices for polysilicon (left, in USD/kg), monocrystalline cells and wafers (middle, USD/$W_p$) and crystalline silicon and thin film modules (right, USD/$W_p$). Data: [8].

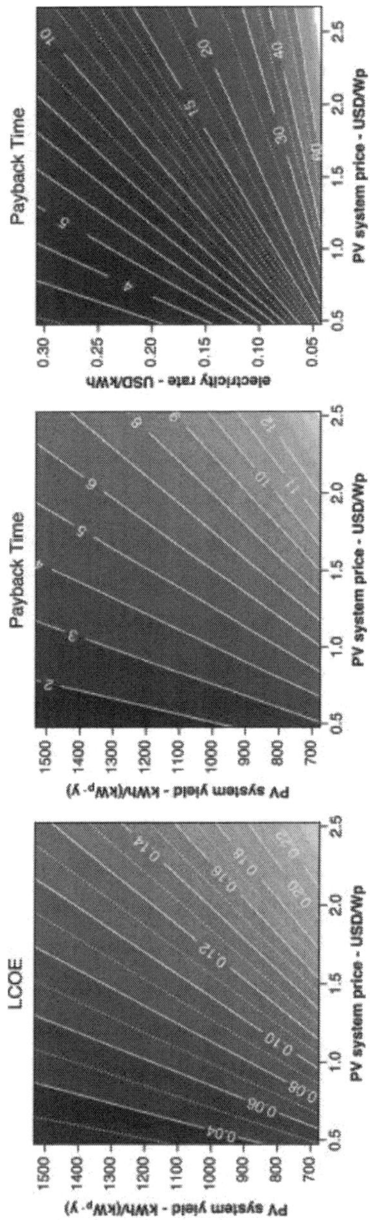


**Figure 2.** Left: effect of PV system price and PV system yield on the levelised cost of electricity from a PV system. Middle: effect of PV system price and PV system annual yield on the investment payback time of a PV system, based on an electricity rate of 0.26 USD/kWh. Right: effect of PV system price and electricity price on the payback-time of a PV system, at a yield of 875 kWh/k $W_p$.

In this paper we present an analysis of the cost structure of PV modules based on five different SHJ cell designs, and, as a reference, conventional diffused junction monocrystalline silicon PV cells. Furthermore, we use the results obtained to evaluate a roadmap towards significant cost reductions in prospective SHJ modules. With the analysis of current production costs, we aim to establish a baseline, but also to compare and rank different SHJ cell designs in terms of cost, and to provide an upfront cost estimation of these different cell designs while they are still under development. Together with previous work [13] focusing on the environmental performance of SHJ solar cells and modules, we aim to perform an ex-ante technological assessment of SHJ technology, and SHJ cell and module production on an industrial scale of designs currently in laboratory stage. To analyze where this technology could go in 10 years, we have also investigated a roadmap towards production of derivatives of the SHJ designs in 2025.

## 2. METHODS

In a previous study we performed a life cycle assessment (LCA) of four of the five SHJ designs studied here, resulting in a detailed description of SHJ cell and module production [13]. LCA studies aim to establish the environmental impact of products, and are based on establishing a life cycle inventory (LCI): a detailed inventory of all material and energy inputs and waste outputs of production of a product. We used the life cycle inventory gathered in our previous study to perform a bottom-up cost analysis using life cycle costing (LCC), by attributing costs to all the material, energy and waste flows in the LCI. Some updates were made to our previous LCI. The LCI used in this study is shown, per processing step, in Table A2, Table A3, Table A4, Table A5, Table A6 and Table A7 in the appendix.

The methodology for LCA is firmly standardised, both in general terms [21] and [22] and for PV specifically [23]. For LCC, this methodology is not yet standardized [24], but in many ways very similar to LCA. We will follow the approach detailed by Rebitzer and Seuring [25] and Hunkeler and Rebitzer [26] for LCC, our previous study followed PV LCA guidelines by Fthenakis et al. [23].

## 2.1. Designs studied

In this study we compared the production of various different SHJ cell designs, and compared this with the production of a conventional crystalline silicon solar cell. The different designs studied here are shown in Fig. 3 and Fig. 4 and described in detail inTable 1 and Table 2. We assumed that production with commercial scale (optimized for throughput) cell processing tools, with wafer throughputs of 3600 wafers/hr (or approximately 120–130 MW/yr dependent on cell efficiency). As in our previous work [13]we assumed that the designs to be based on a 180 μm monocrystalline Cz Si wafer, passivated with intrinsic amorphous silicon (a-Si:H) on both sides, deposited via plasma enhanced chemical vapor deposition (PECVD). The reference design (see Fig. 3) has a standard SHJ structure: an a-Si:H emitter and back surface field (BSF), and an indium-tin-oxide (ITO) transparent conductive oxide (TCO). Metallization is screen-printed silver on both sides. Compared to conventional silicon cells, more silver paste is required for the front grid due to the low-temperature curing process. As shown in [19], there is no single preferred metallization layout in SHJ research and development. Therefore, we have compared SHJ designs with different forms of metallization: the reference design with an Ag grid at the back (Ref-SHJ), and SHJ cells with a PVD deposited Ag layer at the back (PVD-SHJ).

Aside from these designs with only changes in metallization, we have studied three other designs: The Novel Emitter SHJ (NE-SHJ) design has no ITO or a-Si:H layers on the front, but instead has a multifunctional layer which aims to combine the function of passivation, emitter, TCO and antireflective coating in a single layer.

The atomic layer deposition design (ALD-SHJ) replaces the standard SHJ front side with a hole selective $Al_2O_3$/ZnO window layer. The aluminum oxide layer is deposited with ALD, while the zinc oxide layer is sputtered.

The final design studied is an interdigitated back contacted (IBC) SHJ cell (IBC-SHJ). The IBC structure is produced with masked plasma-deposition (PECVD) and -etching, while the contacts are formed at the back by masked PVD of ITO and silver.

We compared these five SHJ designs with a conventional c-Si cell, with a p-type wafer and diffused emitter, aluminum BSF, $SiN_x$ anti-reflective coating (ARC) and fired-through silver grid at the front, and silver soldering pads at the back.

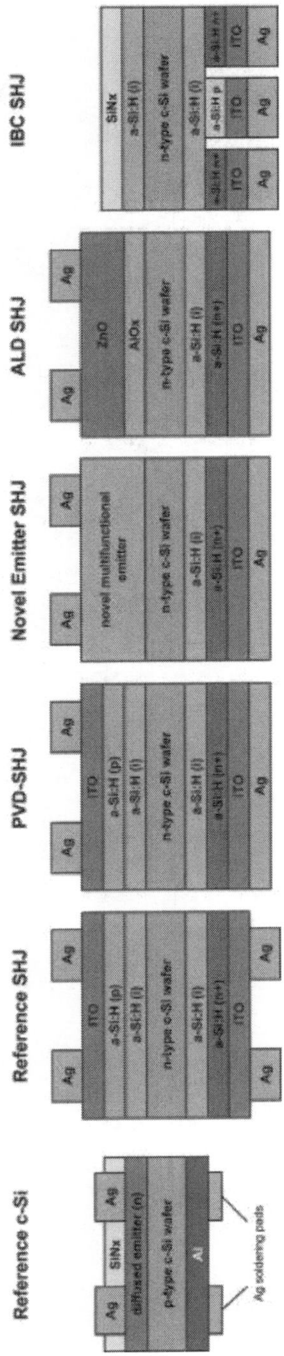


**Figure 3.** Graphical representation of the current scenario cell designs analysed in this study. Based on designs from [13].

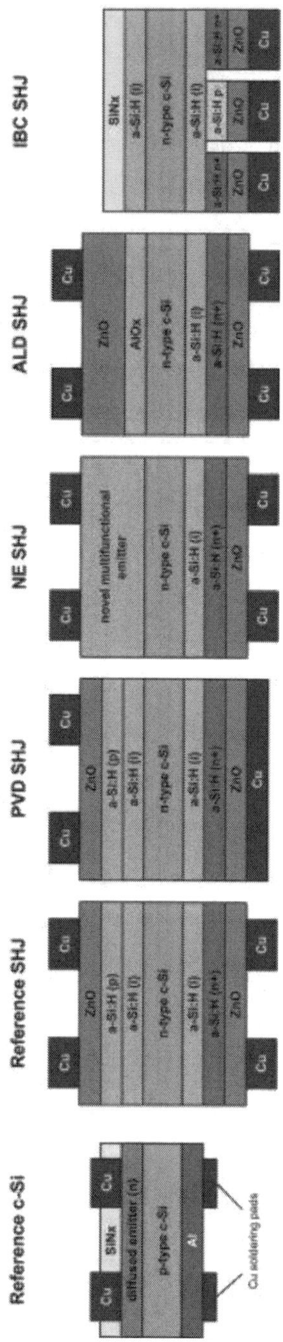


**Figure 4.** Graphical representation of the prospective scenario cell designs analysed in this study. Based on designs from [13].

**Table 1.** Overview of design parameters for the current scenario cell designs. The abbreviations of the design names refer to the discussion of these designs in Section 2.1 and are also shown in Fig. 3 and Fig. 4.

| Short name | Ref-cSi | Ref-SHJ | PVD-SHJ | NE-SHJ | ALD-SHJ | IBC-SHJ |
|---|---|---|---|---|---|---|
| Wafer type | p-type | n-type | n-type | n-type | n-type | n-type |
| Wafer thickness | 180 | 180 | 180 | 180 | 180 | 180 |
| Passivation | | a-Si:H (i) | a-Si:H (i) | a-Si:H (i) | Al2O3 | a-Si:H (i) |
| Emitter | diffused n-dopant | a-Si:H (p) | a-Si:H (p) | multifunctional layer | ZnO | a-Si:H (p) |
| TCO | - | ITO | ITO | ITO (back only) | ITO (back only) | ITO (back only) |
| Metallization front | Ag print | Ag print | Ag print | Ag print | Ag print | - |
| Metallization back | Al print with Ag soldering pads | Ag print | Ag PVD | Ag PVD | Ag PVD | Ag PVD |
| Cell area ($cm^2$) | 239 | 239 | 239 | 239 | 239 | 239 |
| Cell efficiency[a] | 19.4% | 22.4% | 22.7% | 19.7% | 19.5% | 23.3% |
| Module efficiency[a] | 17.1% | 19.7% | 20.0% | 17.3% | 17.2% | 20.5% |

[a] Cell and module efficiencies are monofacial efficiencies.

**Table 2.**Overview of design parameters for the prospective scenario cell designs. The abbreviations of the design names refer to the discussion of these designs in Section 2.1 and are also shown in Fig. 3 and Fig. 4.

| **Short name** | **Ref-cSi** | **Ref-SHJ** | **PVD-SHJ** | **NE-SHJ** | **ALD-SHJ** | **IBC-SHJ** |
|---|---|---|---|---|---|---|
| Wafer type | p-type | n-type | n-type | n-type | n-type | n-type |
| Wafer thickness | **100** | **100** | **100** | **100** | **100** | **100** |
| Passivation | | a-Si:H (i) | a-Si:H (i) | a-Si:H (i) | Al2O3 | a-Si:H (i) |
| Emitter | diffused n-dopant | a-Si:H (p) | a-Si:H (p) | multifunctional layer | ZnO | a-Si:H (p) |
| TCO | - | **ZnO** | **ZnO** | **ZnO** | **ZnO** | **ZnO** |
| Metallization front | **Cu print** | **Cu plate** | **Cu plate** | **Cu plate** | **Cu plate** | - |
| Metallization back | Al print with **Cu soldering pads** | **Cu plate** | **Cu PVD** | **Cu plate** | **Cu plate** | **Cu plate** |
| Cell area ($cm^2$) | 239 | 239 | 239 | 239 | 239 | 239 |
| Cell efficiency[a] | **20.7%** | **25.4%** | **25.3%** | **23.2%** | **23.0%** | **25.9%** |
| Module efficiency[a] | **19.7%** | **24.1%** | **24.0%** | **22.1%** | **21.8%** | **24.6%** |

[a] Cell and module efficiencies are monofacial efficiencies.

### 2.1.1. Cell design efficiencies

This study presents productions cost for five different SHJ cell designs. For most of these designs, it is difficult to assign an accurate cell efficiency, as they are not being produced and tested at a large scale. With an approach similar to that in [18] we have calculated cell efficiencies for each of the SHJ designs, based on an assumed cell efficiency of 22.4% for the reference SHJ design, adjusting the efficiency for each alternative design based on the effect of changes in the cell structure on open-circuit voltage (*Voc*), short-circuit current (*Isc*) and fill factor (*FF*). The assumed efficiencies for each of the SHJ designs, and the assumptions made for their calculations are shown in Table 3.

## 2.2. Scope of the study

The aim of the study reported here was to quantify production costs for the five SHJ cell designs mentioned in the previous section, show a roadmap towards reduction of these costs in the future, and make a comparison with conventional c-Si modules. With the analysis we hope to be able to offer an ex-ante production cost assessment of different SHJ designs currently being developed, in order to be able to steer the R&D towards cost reductions. Furthermore, we hope to identify the importance of SHJ specific design elements in cell and module production to overall costs.

Fig. 7 shows the cell production cycles we have analyzed, which are detailed below in3.1 and 3.2. The focus of our study was on production costs. We therefore established these costs for a functional unit defined as "one Watt of rated (peak) module power output" (USD/$W_p$). The aim of the study is to analyze these production costs based on a breakdown of the production cycles detailed in Fig. 7.

## 2.3. Cost data and calculations

The production costs for the designs studied were calculated by combining material and energy prices with a life cycle inventory (LCI) of cell production. The LCI gives a detailed bill of materials needed for production of the studied solar cells and modules. The LCI was taken from our previous work [13], while cost data was taken from a variety of sources. For an overview of input cost data, see Table A9 in the appendix. Some general cost data assumptions are given in Table 4.

**Table 3.** Overview of estimated SHJ cell performance parameters and the assumptions behind their calculation. Method based on the approach in [18].

| Cell parameter | Ref-SHJ | PVD-SHJ | NE-SHJ | ALD-SHJ | IBC-SHJ |
|---|---|---|---|---|---|
| ***current cell designs*** | | | | | |
| Open-circuit voltage (V) | 0.737 | 0.737 | 0.729<br>– No ITO at front side<br>+ No TCO/a-Si:H interface | 0.715<br>– negative effect of TCO on Voc<br>– front emitter changed | 0.744<br>+ Reduced recombination at front interfaces |
| Short-circuit current density ($mA/cm^2$) | 0.380 | 0.382<br>+ Increased reflectivity backside metal | 0.356<br>+ decreased parasitic absorption of ITO<br>– decreased lateral conductivity emitter/TCO | 0.358<br>– decreased lateral conductivity emitter/TCO | 0.389<br>+ Elimination of shadowing front<br>+ Decreased parasitic TCO absorption front |
| Fill factor | 0.800 | 0.808<br>+Increased conductivity backside metal | 0.760<br>– decreased lateral conductivity emitter/TCO | 0.761<br>– decreased lateral conductivity emitter/TCO | 0.803<br>+ High contact coverage back |
| Cell efficiency | 22.4% | 22.7% | 19.7% | 19.5% | 23.3% |
| Module efficiency | 20.2% | 20.5% | 17.8% | 17.5% | 21.0% |
| ***prospective cell designs*** | | | | | |
| Open-circuit voltage | 0.760 | 0.760 | 0.756 | 0.746 | 0.768 |
| Short-circuit current density | 0.400 | 0.402 | 0.386 | 0.386 | 0.409 |
| Fill factor | 0.820 | 0.828 | 0.796 | 0.797 | 0.826 |
| Cell efficiency | 24.9% | 25.3% | 23.2% | 23.0% | 25.9% |
| Module efficiency | 23.7% | 24.0% | 22.1% | 21.8% | 24.6% |

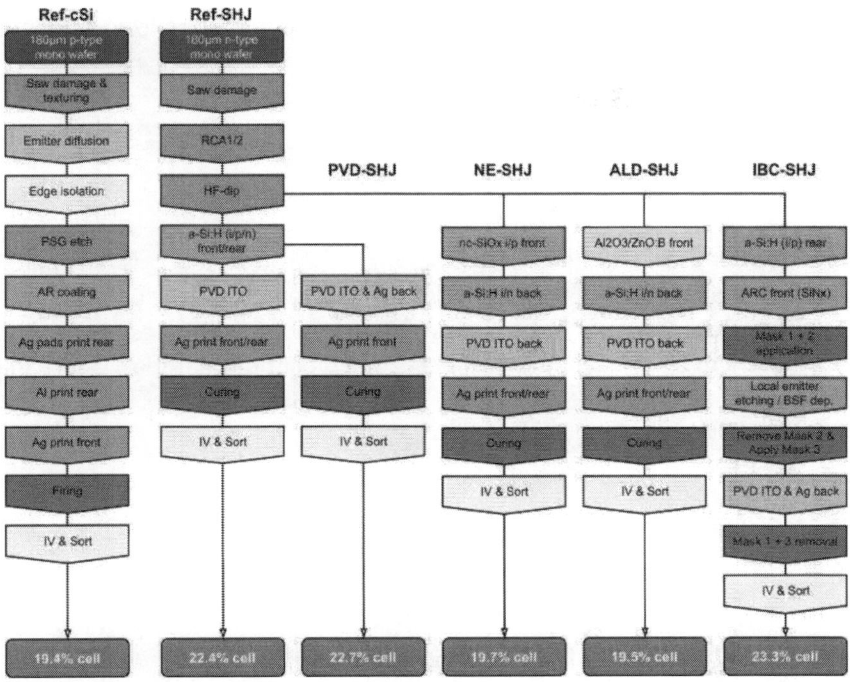


**Figure 7.** Overview of cell processing steps for the different PV cell designs studied here. The process flow for the reference monocrystalline-silicon cell was based on [37].

The LCI data from our previous study [13], was gathered with the aim of analysing the environmental impact of solar cell production. Therefore, it is focused on material and energy in- an outputs, and emissions of harmful substances. As a result, it does not accurately reflect costs at all times. For instance, the price of silver and indium on the global market is not only a result of the materials and energy expended in acquiring them, but is also a reflection of the balance between the supply and demand of these materials. Also, as labor has no direct environmental impact (or one that is negligible), it is also not included in an environmental LCI, at least not on the basis of man-hours.

To account for this difference, we have supplemented the data with processes like labor, and have attributed cost at the highest level in de LCI. For instance, instead of calculating the cost of input energy and materials using life cycle costing (LCC), we have instead used the market price of these inputs. In this sense, our study deviates from a true LCC. For each step in the production chain, the processing costs were calculated as follows:

$$CoP = \frac{\alpha \cdot I_{total} + \sum_{n=1}^{i} (cons_i \times \text{price}_i)}{T_{annual}} \tag{1}$$

where *CoP* is the cost of processing per unit of throughput, $\alpha$ is the capital recovery factor, $I_{total}$ is the total investment expenditure for the processing tool, $T_{annual}$ is the annual throughput of the processing facility, $cons_i$ refers to an annual amount of consumable input (material, labor, and energy inputs), $price_i$ refers to the price per unit of $cons_i$. The capital recovery factor $\alpha$ is defined as:

$$\alpha = \frac{r}{1-(1+r)^{-D}} \tag{2}$$

where *r* is the depreciation rate and *D* is the depreciation period of the processing tool. The capital recovery factor expresses the annual depreciation and interest cost of the capital investment in the processing tool.

Cost data for process inputs was taken from a variety of sources. Equipment prices were taken from market surveys [27], [28], [29], [30] and [31] for cell processing and from [18]for silicon and wafer production. For materials and energy we used data from literature, gathered average market prices, and included globally tradable products at international prices. The data gathered is summarized in Table A7, Table A8 and Table A9 in the appendix.

## 2.4. Sensitivity and uncertainty analysis

The data we have gathered comes from a large variety of different sources, and as such can show large variety in both accuracy, quality and age. To investigate the sensitivity of the overall result to variation in certain parameters, and to analyze the effect of uncertainty in the data on the overall result, we performed a sensitivity and an uncertainty analysis.

**Table 4.** Overview of general economic input parameters. FTE refers to full-time equivalent.

| Parameter | Unit | Value | Source |
|---|---|---|---|
| Depreciation rate | – | 8.0% | [18] |
| Cleanroom cost | $USD/m^2a^{-1}$ | 200 | [18] |
| Skilled labor | $USD/(FTE \cdot a)$ | 70,000 | Own estimate |
| Unskilled labor | $USD/(FTE \cdot a)$ | 50,000 | Own estimate |

For the sensitivity analysis, we varied several parameters over a range defined by either historical data (for consumables, wafer price), or estimated ranges (cell/module efficiency). The resulting recalculated overall results are plotted to show the overall result as a function of change in the studied parameter.

For the uncertainty analysis, we performed a Monte Carlo simulation. From the results and sensitivity analysis parameters were identified that have a large effect on the overall result and have considerable uncertainty. We generated 100,000 random samples of these parameters from log-normal distributions with means of the base values, and sigmas reflecting the uncertainty in each parameter. After that, the results were recalculated for each of the generated samples. The resulting data was presented as boxplots showing the variation in the overall result that was obtained. We opted to use log-normal (instead of normal) distributions as we assumed the parameters under investigation, such as module efficiency and prices, to be more constrained on one side of the mean, and to have a longer tail on the other side, e.g. the uncertainty was assumed to not be symmetrically distributed around the mean. The variance of the lognormal distribution was chosen so the 95th percentile of the distribution reflects the upper level of the uncertainty we want to investigate.

# 3. PRODUCTION COSTS

## 3.1. Silicon, ingot and wafer production

The starting point for all of the devices analyzed in this study is a monocrystalline silicon wafer. Wafer production is generally an activity for dedicated wafer production companies, although supply-demand imbalances in the wafer production chain have resulted in an increased number of PV companies vertically integrating polysilicon, ingot and wafering in their activities [18]. Aside from more vertically integrated firms, the polysilicon production capacity has been expanded significantly since the price of polysilicon peaked in 2011. This has resulted in polysilicon prices dropping sharply, from around 80 USD/kg in 2011 to around 20 USD/kg from the end of 2012 until now (see Fig. 1).

The production chain for polysilicon, ingots and wafers is depicted in Fig. 5. From silica ($SiO_2$), which costs around 20 USD/tonne [32], metallurgical grade silicon (MG-Si) is produced by reducing the silicon-oxide with coke at high temperatures (1900 °C). As this process occurs in an electric-arc furnace, main inputs aside from the $SiO_2$ are electricity, and some form of carbon.

**Figure 5.** Schematic overview of the silicon, ingot and wafer production chain studied here.

This MG-Si has a purity of about 98.5–99.5% [33], while much higher purity (99.99–99.9999%) is required in the solar and especially electronics industry. To achieve these higher purities, the MG-Si is processed into polysilicon, commonly using (hydro)chlorination in a Siemens reactor. A less common alternative is the Fluidized Bed Reactor (FBR), although due to technical challenges this process is not common [18], and is also more capital intensive compared to the Siemens process [34].

In the Czochralski (Cz) process, the polysilicon is "pulled" into monocrystalline silicon ingots, which are sawn into wafers with wire saws. In this process, silicon is lost as kerf loss, due to the abrasive sawing of the ingots into wafers. The silicon usage per wafer is reduced by decreasing wafer thickness and kerf losses. Current wafers are generally 180 μm thick [35], and sawing losses amount to 130 μm of silicon per wafer [18].

For material and energy requirements in polysilicon, ingot and wafer production, we used a life cycle inventory (LCI) from the ecoinvent database [33]. This database analyses the full life cycle of production of a 270 μm Cz wafer. The data from this inventory were updated to account for the different wafer thicknesses we investigated. Furthermore, as ecoinvent data does not accurately reflect the cost of capital expenditures, we based equipment and facility costs on Goodrich et al. [18], who studied the polysilicon-ingot-wafer production chain as a basis for cost roadmaps for various types of c-Si based PV modules.

By attributing costs to all the material and energy inputs from the LCI we calculated a cost structure for monocrystalline silicon wafers, from silica sand to wafer sawing. Table 5shows the main input parameters for Cz ingot production. Fig. 6 shows the calculated cost structures for wafers of various thicknesses. The 100 and 80 μm wafers were assumed to be sawn with diamond wire saws (see Table 6).

**Table 5.** Main input parameters for Cz ingot production. Amounts per kg of Cz Silicon ingot produced.

| Input | Amount | Unit | Source | Unit cost | Source |
|---|---|---|---|---|---|
| PV grade silicon | 1.07 | kg | [33] | 16.47 | Own model |
| Electricity (hydro) | 85.6 | kWh | [33] | 0.025 | [18] |
| Argon | 5.8 | kg | [33] | 0.714 | Own survey |

## 3.2. Cell processing

The analyzed process flows are shown in Fig. 7. The basis for all designs studied is an 180 μm monocrystalline silicon wafer, n-type for the SHJ and p-type for the c-Si reference design. For simplicity, wafer costs are assumed equal for p-type and n-type wafers. The following section discuss the cell processing flow for each of the studied designs.

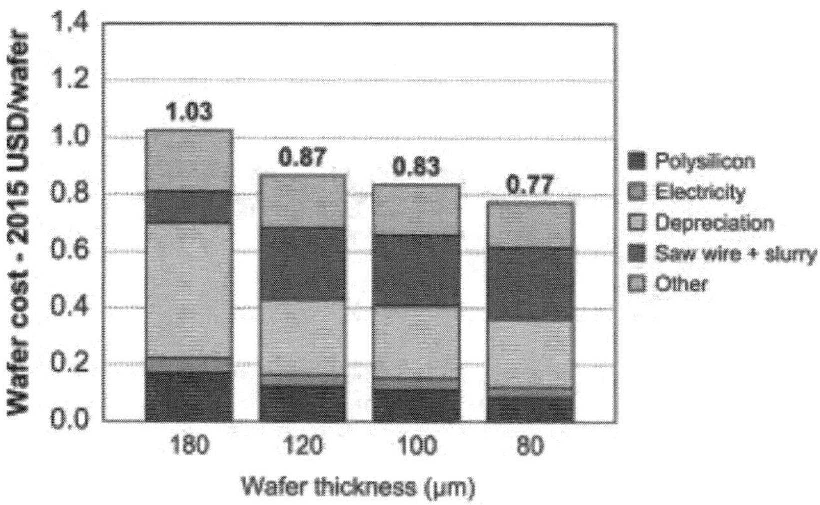


**Figure 6.** Overview of wafer production costs for various wafer thicknesses. In this study, p-type and n-type wafers are assumed to have equal production costs.

### 3.2.1. Conventional *p*-type c-Si cell

The p-type c-Si cell is assumed to be produced according to a standard process flow. Wafers are treated for saw damage and textured, prior to emitter diffusion in a diffusion furnace. After diffusion, a wet chemical treatment step is performed for edge isolation and phosphosilicate glass (PSG) removal. The silicon nitride (SiN*x*) ARC is deposited with PECVD. Metallization is screen printed (Al with Ag soldering pads on the back, Ag grid on the front) and fired in a firing furnace at high temperature.

### 3.2.2. Reference SHJ cell

A reference case for SHJ cells was analyzed based on a conventional SHJ cell structure. The process flow for this Reference SHJ design (Ref-SHJ) is shown in Fig. 7, while the cell design is described in Table 1 and Fig. 3. The Ref-SHJ cell has a symmetric layout, with wafer, intrinsic a-Si:H, doped a-Si:H, ITO and metallization.

**Table 6.** Main input parameters for wafer production.

| Input per m$^2$ | wafer thickness | | | | Unit | Unit cost (USD) | Source |
|---|---|---|---|---|---|---|---|
| | 180 | 120 | 100 | 80 | | | |
| Cz silicon | 0.74 | 0.54 | 0.49 | 0.38 | kg | | Own calculations based on [33] |
| Electricity | 8.0 | 8.0 | 8.0 | 8.0 | kWh | | [33], assumed constant for varying wafer thickness |
| SiC fluid | 1.2 | 0.7 | | | L | | [18,33] |
| - recycled | 3.5 | 2.2 | | | L | | [33] |
| Cutting fluid | | | 3.0 | 2.0 | L | | Assumption, cost from [18] |
| Steel saw wire | 5.0 | 5.0 | | | km | 0.5 | [34] |
| Diamond saw wire | | | 83.0 | 83.0 | m | 0.125 | Estimate based on [36] |

To treat the wafers for saw damage, texture it to achieve better light-trapping, and remove oxides from the surface of the wafers, the first process in the cell processing is at wet chemical treatment. In our model, the wafers are treated with (1) Sodium hydroxide for saw damage removal, (2) A two step RCA cleaning procedure, and (3) a HF dip for oxide removal and texturing.

In terms of costs, this process main costs are consumables (the etchant solutions), capital expenditures (CapEx) and waste treatment. A survey on wet chemical treatment tools [27] shows average CapEx to be 293±52 USD/(wafer/h), resulting in a wafer processing cost of 0.017 USD/wafer for capital investments, while total cost-of-ownership (CoO) including consumables is 0.075 USD per wafer. The etchants used in this process account for over 30% of these costs.

After wet chemical treatment, the electronic structure of the SHJ cell is applied by deposition of thin a-Si:H layers on both sides of the wafers. On one side, an a-Si:H (i) layer is deposited, followed by a p-type a-Si:H layer. On the other side, an $i/n^+$ a-Si:H stack is deposited. Main consumables are electricity, silane, hydrogen, and water. Because of the very thin layers applied (5 nm for the intrinsic layers, up to 20 nm for the doped layers) the material consumption is very low. As a result, most of the cost of this process step results from the capital expenditures for the processing tools. CapEx for PECVD was based on equipment costs for PECVD tools used for a-Si:H deposition in the preparation of thin film modules. A survey [30] showed large variation in the cost of these tools (ranging from 1489 to 5393 kUSD/(wafer/h), especially the older tools were found to have very high costs. PECVD tools for deposition of SiN*x* ARC layers were found to have much lower costs, on average 882 USD/(wafer/h) [31], but as this process occurs at higher pressures, we assume these tools to be cheaper compared to those for a-Si:H deposition. We have selected the two most recent tools from the survey [30] to calculate PECVD costs for a-Si:H deposition, and found these to be 1676 USD/(wafer/h). Resulting processing costs were found to be 0.050 USD/wafer for capital expenses, or 0.078 USD/wafer including consumables and utilities.

PECVD is followed by the deposition of a TCO layer, in this case ITO, on both sides of the wafer. This transparent conductive oxide-layer is applied to improve lateral conduction in the device, while maintaining a high absorbance of light in the p/n region of the device. This deposition of ITO is commonly assumed to be expensive, as indium is a scarce and quite expensive material (750 USD/kg). According to our cost model however, this processing step is not as expensive as expected. The very thin layers deposited do not require much material (0.028 gram/wafer), and utilization factors in the sputtering tools are quite high (88%, resulting in ITO consumption of 0.032 gram per wafer). The cost of ITO target material was calculated to be about 0.85 USD/gram, although this is a rough estimation, and likely an overestimation, as we assume that commercial targets to approach raw materials prices (0.75 USD/gram). Capital costs for the sputtering tools we reviewed were found to be on average 1573 USD/(wafer/h) [28]. The total CoO of TCO deposition was found to be 0.087 USD/wafer, of which 0.027 USD/wafer (31%) was due to the consumption of ITO.

### 3.2.3. Silver-based metallization

A main cost factor in cell processing, as expected, was found to be the silver based metallization. As SHJ solar cells have to be processed at low temperatures (below 250 °C), there is a need for low-temperature silver paste. This paste is more expensive than conventional metallization paste, and more of it is needed to reach the required contact resistance values in the cells. We assume that the higher price of low temperature paste is mainly the result of a higher silver content in these pastes, but supply and demand dynamics possibly also play a role, as the market for low-temperature paste is much smaller to that of regular (high-temperature) paste.

Current silver paste cost were conservatively estimated at 820 USD/kg for high-temperature and 1060 USD/kg for low-temperature paste. Because of the lower conductivity of the cured low-temperature paste, we assume that double the amount of silver is required compared to high-temperature silver paste metallization. A metallization grid is therefore assumed to require 200 mg silver per side of the wafer, or 250 mg of paste, assuming 80% loading of the paste with silver. For the reference SHJ design, we assume that the application of a silver grid on both sides of the solar cell, thus the total paste requirement per cell is 0.5 grams. Total materials costs for metallization amount to 0.55 USD/wafer, or 0.10 USD/$W_p$. In current and past SHJ cell R&D many different metallization layouts have been studied for the backside of the cell, including silver screen printed grids, aluminium PVD back contact and silver PVD back contact. An aluminum backside would have the lowest material costs, but impacts the efficiency of SHJ cells severely as it is shown to result in a poor internal quantum efficiency in the infrared region of the spectrum without a more complex rear structure [38].

### 3.2.4. Silver PVD SHJ cell (PVD-SHJ)

The silver PVD SHJ cell (PVD-SHJ) is another design strongly based on the reference SHJ cell. Here, instead of a bifacial Ag grid, the backside contact is formed from PVD deposited silver. The cost of silver target material is lower compared to silver paste, resulting in a material cost decrease of 0.26 USD/wafer, to a total of 0.27 USD/wafer for metallization materials (0.059 USD/$W_p$). Here, we assumed that the thickness of the PVD deposited layer to be 0.2 µm[38], while the utilization rate of the PVD tool is assumed to be 74% [28], resulting in a silver consumption of 0.07 g/wafer for the rear side metallization. Capital costs do increase by 0.04 USD/wafer due to the PVD tool, but the resulting costs for metallization are still considerably lower at 0.33 USD/wafer compared to 0.55 USD/wafer for the bifacial silver grid.

### 3.2.5. Novel emitter SHJ cell

The novel emitter design is an SHJ design that focuses on replacing the scarce and expensive ITO with abundant materials, at the same time simplifying the processing sequence by cutting out on process, namely TCO deposition. The process flow for this design is shown in Fig. 7. As we can see from this figure, the basis of this design is very similar to the reference design. The changes occur however during the PECVD step. Instead of a p-doped a-Si:H layer, a transparent p-type conductive PECVD layer is deposited at the front side. The layer is deposited to function as both an emitter and a transparent conductive layer.

The replacement of the ITO layer on the front results in an increased requirement for PECVD. This does however not result in a strong decrease of throughput of the PECVD step as the p-layer is replaced by the combined layer and the main time-consuming factors are a result of creating and releasing the vacuum in the processing chambers, which is done once less. For the deposition of a total of 130 nm layer thickness, which is an increase in terms of total deposited layer thickness by a factor of 2.6, throughput drops with only about 4.5% compared to the reference design, based on throughput figures for a the selected PECVD tools mentioned in Section 3.2.2. This results in an increase of processing cost for PECVD by 0.025 USD/wafer to a total of 0.10 USD/wafer. The absence of the ITO deposition on the front however gives an advantage of 0.03 USD/wafer, largely due to increased throughput and an ITO consumption cost reduction of 0.013 USD/wafer.

### 3.2.6. ALD aluminum-oxide/zinc oxide SHJ cell

The ALD aluminum-oxide/zinc-oxide SHJ cell design (see Fig. 7) is an R&D design that tries to address two issues with conventional SHJ cells, namely (1) ITO replacement for more abundant ZnO and (2) introduction of a thin-film deposition process that is less energy-intensive compared to PECVD and does not require operation at vacuum. The back-side of the cell is similar to that of the Reference SHJ and Novel Emitter SHJ design, with an i/n-stack of a-Si:H deposited with PECVD. On the front side, an $Al_2O_3$passivation layer and ZnO window-layer are deposited with ALD and sputtering, respectively. Alternatively, this process could also involve $MoO_3$ as a high-workfunction hole-selective contact.

Average CapEx for the ALD tools surveyed [39] was found to be 796 USD/(wafer/h). As consumable consumption is low, the capital costs account for over 60% of the cost of processing a wafer with ALD. Another major factor is the cost of trimethylaluminum, which at 845 USD/kg accounts for over 20% of the cost of processing. Energy consumption in this atmospheric pressure process was found to be much lower compared to PECVD.

The relative cost of the PECVD deposition of the back-side i/n a-Si:H stack (25 nm total thickness) is quite high, as the throughput of the PECVD tool only increases by about 20% on average, while the capital related cost of processing only decreases by 11% compared to deposition of 50 nm of total a-Si:H layer thickness. As this cost advantage is minimal, the total cost of processing the ALD cell is slightly higher compared to that of the reference SHJ cell.

### 3.2.7. IBC-SHJ cell

The interdigitated back-contacted heterojunction (IBC-SHJ) cell is quite different from the reference design in terms of cell and module structure. The device has an assymetric structure, with interdigitated emitter and back-surface field and corresponding contacts. For this design we assume this interdigitated structure to be applied by using a masked PECVD deposition and etching process, based on a process described [40]. The process flow of this cell design is shown in Fig. 7, and detailed below.

The IBC-SHJ cell undergoes the same wet chemical treatment as the other designs, however, after this step the differences are pronounced. First with PECVD, on the front side of the cell a passivation layer (intrinsic a-Si:H) and $SiN_x$ antireflective coating (ARC) are applied, while on the backside an a-Si:H emitter is applied. Then, a base mask is applied, followed by the mask for local plasma etching of the emitter and deposition of the a-Si:H backsurface-field (BSF). The BSF mask is removed and replaced with a second mask for TCO deposition and deposition of the evaporated silver back-contact. We have assumed that contact coverage at the rear side to be 80%, with a thickness of 0.4 micron, thus requiring 140 mg of silver target per wafer, assuming a 74% utilization rate in the sputtering tool.

We have assumed silver to be required for this IBC contact, as the reduced contact area would result in higher contact resistance when using aluminium.

The cost structure of production of this design is affected in three ways: (1) more processing steps are required, leading to a higher cost of capital, (2) replacement of the silver front grid with a silver PVD back-contact increases material costs for metallization, more silver is required for the back-contact, (3) the application and removal of masks increases handling, yield losses and material requirements.

To calculate the cost of producing the IBC structure, we have made several (rough) assumptions: (1) each process requiring a mask application or removal has a 10% lower throughput, (2) a mask application or removal step induces a 2.5% yield penalty, and (3) mask costs were estimated to be 0.05 USD/wafer.

The increased requirement of PECVD, coupled with the application and removal of masks, requires a three-step PECVD. This leads to a decreased throughput, and as a result, a strong increase in capital costs. Furthermore, material requirements are much higher, and yield losses increase due to the number of handling steps when applying and removing the masks. The total cost of processing of PECVD therefore increases more than three-fold compared to the reference design, to 0.27 USD/wafer, of which 0.03 USD/wafer due to yield losses (compared to only 0.0011 USD/wafer for the reference SHJ design).

The replacement of the front-side silver grid made with low-temperature silver paste with an interdigitated PVD silver contact decreases the cost of metallization by 0.33 USD/wafer compared to the Ref-SHJ design. This is mainly the result of the substitution of low-temperature silver paste screen printed grid with a silver PVD metallization structure. Per watt-peak, metallization cost drop by 0.06 USD. Overall, CapEx costs increase by over 0.2 USD/wafer, but, due to the absence of low-temperature silver paste, consumable costs decrease by over 0.3 USD/wafer, resulting in processing costs that are 0.019 USD/wafer lower compared to the reference design.

## 3.3. Modules

For the current designs (except the IBC design), SHJ module production is quite similar to that of regular crystalline silicon modules, with the exception that the soldering step cannot be performed for SHJ cells, as this would lead to too high cell processing temperatures [19], [41] and [42]. Therefore, the stringing and tabbing is performed with conductive adhesive (CA), which is cured during the module lamination step [41]. For the IBC-SHJ designs, the 1.8 grams/module of CA is printed as dot-contacts, while for the other SHJ designs, 3.6 grams/module of CA is applied on the busbars.

Before module production, cells are tested and sorted according to their IV-characteristics, so that they can be matched in 60-cell modules for an optimal power rating. Cells have to be handled to be layed up, tabbed and stringed and laminated into the module, in a sandwich of glass, backsheet and EVA. We reviewed CapEx for module production equipment [43], [44], [45] and [46] and found total costs for a module production line to be 4.8±0.7 million USD for a line with a throughput of 60 modules/hr, resulting in a CapEx cost of 0.006–0.007 USD/$W_p$.

Conductive adhesive is more expensive than regular solder, mainly because silver is often used as a conductive material (as opposed to cheaper tin and lead). At the same time, the resistivity of conductive adhesive compared to regular solder is a concern [41]. For the performance stability of SHJ modules and protection against moisture related degradation, we assume the backsheet to be a combination of a standard Tedlar ®-Polyester-Polyamide (TPA) backsheet and an aluminium foil moisture barrier. Table 7lists the costs of various module components assumed in this study. Prices for glass, frame, backsheet and ethyl-vinyl-acetate (EVA) where estimated based on average market prices.

**Table 7.** Overview of module component costs. Costs specified in USD/module relate to modules of standard size (1.65 m × 0.99 m).

| Component | Cost | Unit | Source |
|---|---|---|---|
| Frame | 13.5 | USD/module | Market survey |
| Glass | 5.50 | USD/$m^2$ | Market survey |
| Standard backsheet | 5.04 | USD/$m^2$ | Market survey |
| SHJ backsheet with Al foil | 7.00 | USD/$m^2$ | Own assumption |
| IBC backsheet with Cu foil | 10.0 | USD/$m^2$ | [47] |
| EVA | 1.85 | USD/$m^2$ | Market survey |
| J-Box | 6.50 | USD/module | [18] |
| Stringing/Tabbing | 2.50 | USD/module | [18] |
| Conductive adhesive | 750 | USD/kg | [47] |

The module design for the IBC-SHJ design is quite different. The interdigitated contacts mean that regular tabbing and stringing cannot be employed for cell interconnection. Rather, the cells are interconnected by glueing them with conductive adhesive to a backsheet/foil with prestructured copper contacts on it. The conductive adhesive is printed onto the foil in small dots. The EVA foil on the backside is patterned, to make holes at the points where the conductive adhesive is printed. The sandwich of glass, EVA, cells, patterned EVA, conductive adhesive, and backsheet/foil is laminated in one step in which the conductive adhesive is simultaneously cured. Table 8 summarizes the differences between the three module layouts (Table 9).

**Table 8.** Overview of difference in 60 cell module design and production between c-Si and SHJ modules.

| Module element | c-Si | SHJ | IBC-SHJ |
|---|---|---|---|
| Stringing | Soldering tabber/ stringer | CA & tabber/ stringer[a] | CA & lamination[b] |
| Backsheet | Standard TPA | TPA with aluminium foil[c] | TP with structured Cu foil |

[a] Cell stringing is assumed to be performed with a conventional tabber/stringer, but with conductive adhesive (CA) instead of regular solder.

[b] For IBC modules, we assume cell interconnection to be made during lamination, when conductive adhesive (CA) that was printed onto the structured foil is cured together with the EVA sandwich.

[c] SHJ modules are assumed to require a backsheet that is more resistant to moisture ingress. We assume this backsheet to be composed of a standard TPA backsheet with an additional aluminium foil layer.

Fig. 10 shows a breakdown of module costs. Glass, backsheet and frame contribute most substantially to the overall module costs, while the addition of conductive adhesive results in a cost penalty of 0.007–0.014 USD/$W_p$ for the SHJ designs. The structured backsheet for the IBC design increases the module cost by about 0.06 USD/$W_p$.

## 4. PROSPECTIVE DESIGNS

The results for current designs indicate, as expected, main contributions for wafer and metallization to overall cell production cost. Other significant factors are PECVD and TCO sputtering while wet pretreatment of wafers contributes very little to overall production costs. In this paragraph we will detail, per processing step, a roadmap towards possible reduction of cell production costs. For this section we assume production in 2025.

**Table 9.** Effect of design changes on module production costs.

| Improvement | Effect on production costs (USD/$W_p$) | | | | | |
|---|---|---|---|---|---|---|
| | **Ref-cSi** | **Ref-SHJ** | **PVD-SHJ** | **NE-SHJ** | **ALD-SHJ** | **IBC-SHJ** |
| **Efficiency** | −0.051 | −0.060 | −0.064 | −0.088 | −0.089 | −0.071 |
| Increased cell efficiency | −0.048 | −0.057 | −0.060 | −0.083 | −0.084 | −0.067 |
| Increased cell to module power ratio | −0.003 | −0.003 | −0.004 | −0.005 | −0.005 | −0.004 |
| **Wafer** | −0.056 | −0.054 | −0.042 | −0.051 | −0.054 | −0.046 |
| **Metallization** | −0.046 | −0.093 | −0.050 | −0.064 | −0.064 | −0.034 |
| Ag paste reduction | −0.018 | −0.060 | −0.029 | −0.034 | −0.034 | 0.000 |
| Ag replacement with Cu | −0.029 | −0.033 | −0.020 | −0.030 | −0.030 | −0.034 |
| **ITO replacement with ZnO** | 0.000 | −0.010 | −0.010 | −0.008 | −0.009 | −0.010 |
| **Module Layout** | −0.020 | −0.025 | −0.017 | −0.028 | −0.029 | −0.017 |
| Frame cost reduction | −0.020 | −0.017 | −0.017 | −0.020 | −0.020 | −0.017 |
| Bifacial module | | −0.008 | | −0.009 | −0.009 | |
| **Total** | −0.174 | −0.234 | −0.182 | −0.230 | −0.235 | −0.178 |

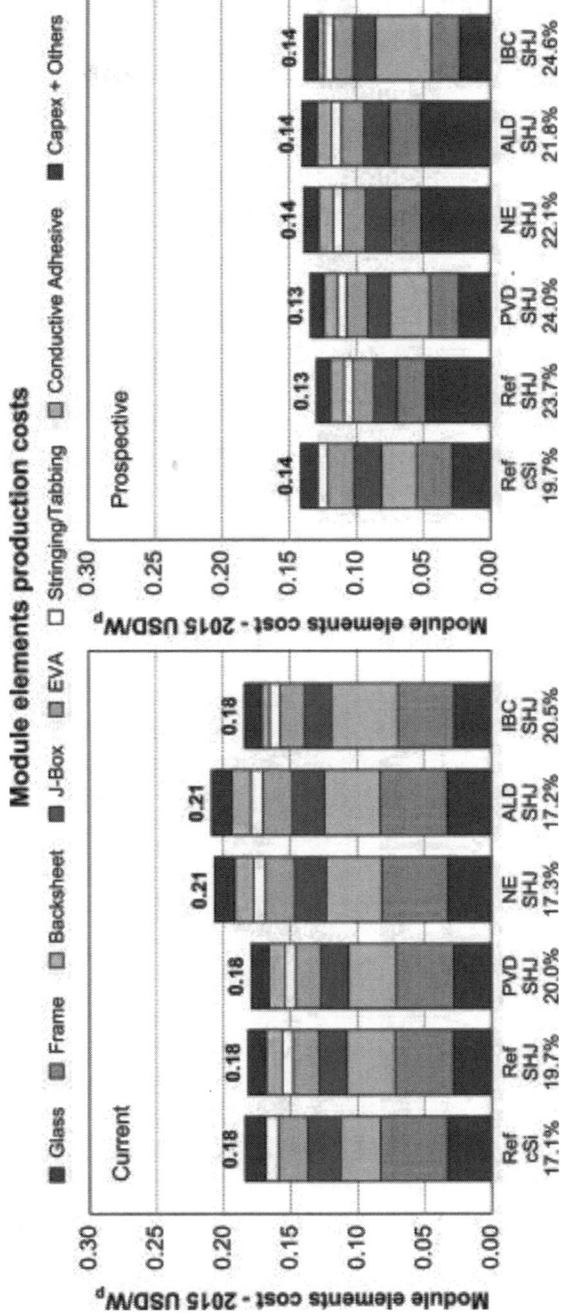


**Figure 10.** Breakdown of the cost of module elements. Totals are indicated in bold above the bars. The percentages below the labels indicate module efficiencies.

## 4.1. Wafer costs

Logically, wafer prices are mainly influenced by two parameters: silicon usage per wafer, and silicon price per kg. Apart from a peak in the silicon price in 2008, silicon prices are relatively stable between 15 and 30 USD/kg (see Fig. 5). Cost reductions for silicon wafers therefore mainly focus on reducing wafer thickness and sawing losses. Goodrich et al. [18] project wafer prices to decrease significantly, due to these two cost reduction strategies, but also include the change from standard to diamond-wire saws, as this should increase throughput [35] and [48]. Diamond wire saws furthermore are more durable and do not require silicon-carbide slurry but work with a much cheaper cutting fluid based on water and surfactant [18].

Regular tabbing-and-stringing of cells into modules puts the solar cells under significant amounts of mechanical stress. With decreasing wafer thickness, this mechanical stress increases up to a point at which the cost reductions achieved by thinning the wafers is matched by the added cost of production yield losses. Therefore, we assume a minimum wafer thickness of $100\ \mu m$ for all prospective modules.

In silicon production, a gradual shift from Siemens to Fluidized Bed Reactor (FBR) process that is already started is expected to continue, leading to a FBR market-share of just over 30% in 2025 [48]. FBR is currently a new technology and as discussed inSection 3.1, is quite complex. Cost reduction with this technology is therefore not currently expected [18]. Therefore, we modeled prospective silicon costs based on continued use of the Siemens process.

As shown in Fig. 6, even with conservative assumptions regarding silicon and wafer production, wafer cost can decrease substantially from the current modeled cost of 1.03 USD/wafer to 0.83 USD for diamond-wire sawn $100\ \mu m$ wafers and 0.77 USD for $80\ \mu m$ diamond-wire sawn wafers. However, it is not expected that module technology will be compatible with wafers below $100\ \mu m$ by 2025, especially for monocrystalline cell based modules [48]. Therefore, we have based our prospective modules on wafers of $100\ \mu m$, for all cell designs. Although the low-temperature processing of SHJ cells is more compatible with thinner wafers than the higher temperature processing of standard c-Si cells, we have not assumed different wafer thicknesses to avoid bias in cost due to different wafer thicknesses.

## 4.2. Metallization

Cost reductions in solar cell metallization focus on two aspects: (1) reduction of material consumption and (2) substitution of high-cost metallization materials. Because of the high cost of silver-based metallization, the amount of silver used per cell has decreased, and is expected to continue decreasing for the coming years [35]. Although silver use per cell is constrained by minimum conductivity requirements [49], it is expected to decrease from 100 mg per cell in 2015 to about 40 mg/cell in 2025 [48]. This material use reduction alone results in significant cost reductions if application to all designs would be possible. For the conventional mono-crystalline cell, reduction from 100 to 40 mg/cell of silver would drop metallization costs by 36%, or 0.02 USD/$W_p$. A reduction to 40 mg/cell of silver would actually mean that the Al-back paste would contribute the majority of metallization costs. However, as the Al paste is essential for the Al-BSF in the Ref-cSi design, the backside metallization of this design remains the same.

As the silver related metallization costs are much higher for the current SHJ designs (aside from the IBC design), the effect of this strong reduction in silver use per cell is very substantial. For the SHJ designs, metallization costs drop by 58% (Ref-SHJ), 49% (PVD-SHJ) and 45% (NE-SHJ and ALD-SHJ) or by 0.03 to 0.06 USD/$W_p$.

Further cost reductions can be obtained by replacing silver with copper. Copper has similar conductivity at only a fraction of the cost of silver. Currently, the cost of copper is less than 2% of the cost of silver. However, copper oxidizes much more easily compared to silver, and diffusion of copper into the silicon substrate is also an issue, as copper atoms diffusing into the silicon can negatively affect performance [50], through a variety of mechanisms [51]. Copper diffusion can be prevented by a nickel barrier layer, while oxidation is prohibited by covering the copper with a thin layer of silver applied with electroplating. Another approach to reduce copper oxidation is to dry/cure the applied metal contact lines in an inert (nitrogen) atmosphere [52]. Printed copper busbars that were printed on top of a $SiN_x$ ARC and cured at low temperature in a $< 10$ ppm oxygen atmosphere were shown to result in high module efficiencies, while subsequent thermal cycling and damp heat test showed no copper diffusion into the silicon [50]. Other recent results have furthermore demonstrated a newly developed copper paste that offers similarly low contact resistance compared to silver-based pastes, and can be cured at low temperature in "normal" atmosphere after screen printing [53].

However, line resistances were found to be much higher compared to both low temperature silver paste and high temperature silver paste [53].

In the area of copper metallization, copper electroplating is often mentioned as the prime candidate, as it has been shown to offer the possibility of high-efficiency, silver-free heterojunction solar cells, at a large cost-advantage compared to silver screen printed SHJ cells [54] and [55]. However, electroplated contacts require a multi-step process, consisting of screen-printing and curing a plating resist ink [18], plating different metal layers, and stripping the resist ink in a wet chemical tool. Despite this added complexity, the cost of plating is indeed significantly decreased compared to silver screen-printed contacts [55].

Although adoption of copper metallization in mass production of PV is not expected before 2018 [35], we have modelled our prospective designs with copper plated contacts, as we feel silver substitution is not only advisable for cost reductions, but also necessary, as silver is becoming an increasingly scarce material [56], [57], [49] and [58], so an upwards trend of the silver price is likely. The copper plated metallization assume a metal stack of nickel, as a seed layer and copper diffusion barrier, copper, as the main contact metal, and silver, as a capping layer which prevents oxidation of the copper. Electrolyte costs were modelled based on their composition, with the cost of plating equipment was taken from [55].

The cost reductions resulting from substitution of silver with copper result in additional cost savings of 0.01 USD/$W_p$ for the Ref-cSi design, 0.03 USD/$W_p$ for the Ref-SHJ design and 0.02 USD/$W_p$ for the other SHJ designs.

## 4.3. TCO sputtering

Especially for SHJ solar cells, Indium-Tin-Oxide (ITO) is a commonly used TCO material. Because of the high cost and perceived scarcity of Indium, replacement of ITO with other materials is under investigation in many R&D projects. Zinc oxide, doped with boron or aluminium is a commonly named alternative [18]. Our results show that the deposition of ITO layers does not contribute much to the overall costs of cell and module production, as the amount of material used is very low. Still, we assume our prospective designs to be ITO-free, as there are concerns about the availability of indium [59] and [60] and ZnO targets were calculated to be slightly cheaper. Due to the higher resistance of ZnO compared to ITO, this replacement does however result in an efficiency penalty, due to an increased series resistance (*Rs*). Also, the workfunction of ZnO is sub-optimal for contacting a p-type a-Si:H layer. Aside from materials, TCO sputtering seems to be a mature process, with relatively low capital costs and high throughput.

## 4.4. Modules

Module design and materials have been largely the same for quite some time, and it is expected that the market shares of alternative module designs will remain limited up until 2025 [35] and [48]. Material use is expected to slightly decrease in the coming years, due to a step from 3.2 to 2 mm thick glass, decreasing EVA and backsheet thickness, and reduced aluminium usage in module frames [35]. For the aluminium frame we assume that the cost is reduced proportionally to the mass decrease shown in [35]. For the other materials, accurate data could not be found. Increased light transmission and trapping for the modules results in a slight increase of cell-to-module power ratio [35].

On the other hand, because of the high price of silver, the silver content of conductive adhesive will likely decrease further. Another development is the increasing attention for frameless and/or bifacial modules. In these modules, the backsheet is replaced with another pane of glass, and, possibly, the frame is also omitted.

The Ref-SHJ, NE-SHJ and ALD-SHJ designs in our prospective study are bifacial by design, thus, their prospective module layouts are assumed to be bifacial, replacing the backsheet with a pane of glass. For the Ref-cSi design and the other SHJ designs, we assume a standard module layout, with a backsheet. We assume all modules to have an aluminium frame, which is, as discussed above, decreased in weight and cost.

# 5. CURRENT AND PROSPECTIVE CELL AND MODULE PRODUCTION COSTS

Cell production costs (in USD/$W_p$) are shown in Fig. 8. As expected, a main contributor to cell production costs is the wafer, for all designs. The SHJ designs have cell production costs ranging from 0.31 to 0.35 USD/$W_p$, while the cell production cost for the c-Si cell was found to be 0.31 USD/$W_p$. The IBC design benefits strongly from its high efficiency and substitution of silver paste with PVD silver, which offsets the cost penalties for lower yield and throughput, while the bifacial Ref-SHJ cell shows a strong cost-penalty for requiring a large amount of low-temperature silver paste. The novel emitter and ALD SHJ designs are affected by their lower efficiency, and thus have higher production costs compared to the reference design. The PVD-SHJ design has the lowest cost of the SHJ designs, as it benefits from high efficiency and substitution of low-temperature silver paste.

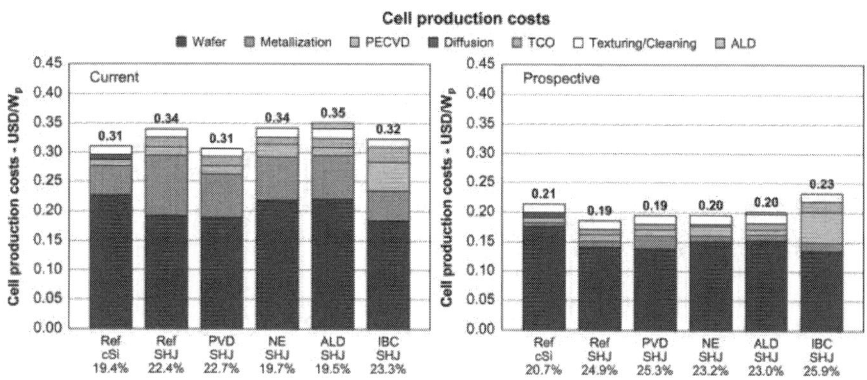


**Figure 8.** Overview of cell production costs for the five silicon heterojunction designs and a conventional monocrystalline silicon device. Left: current production costs; Right: prospective production costs. The percentages below the labels indicate cell efficiencies.

Comparing the Ref-cSi cell with both the Ref-SHJ and PVD-SHJ designs shows that the choice of metallization layout likely determines the competitiveness of SHJ technology. There is a strong cost penalty for using silver paste on both sides of the SHJ cell, that cannot be offset by the increase in efficiency.

When intercomparing the SHJ designs, we see that the replacement of the front ITO in the NE-SHJ design does not result in a cost reduction compared to the Ref-SHJ design, as there is a significant assumed efficiency penalty for this novel SHJ design. For the ALD-SHJ design, this effect is also shown.

As indium is a relatively expensive material, the effect of replacing the ITO layer was expected to be larger. However, the LCA model [13] already showed that only very small amounts of indium are required as the layer thickness is very small. For two 80 nm layers of ITO, slightly over 0.03 grams of ITO was calculated to be required per cell (based on typical utilization rates, density of ITO, and the volume of the layer). The increased requirement for PECVD in the novel emitter design slightly decreases the advantage obtained by having no sputtered TCO layer, as the increased a-Si layer thickness leads to a slightly decreased throughput, and thus an increased cost of processing (*CoP*, see Eq. (1)). The main determinant for higher production costs of the NE-SHJ design is however the decreased efficiency compared to the Ref-SHJ design.

## 5.1. Module costs

Fig. 9 shows the costs of complete PV modules based on the designs studied, while Fig. 10 shows the cost of the module materials. The requirement of conductive adhesive results in a small cost penalty for SHJ modules, while the structured back foil required for the IBC-SHJ design results in a larger cost increase for this particular module. These added costs compared to conventional modules are not completely be balanced by the increased module efficiencies. Especially for the NE-SHJ and ALD-SHJ designs, the relative increase in efficiency is to small to offset added costs of cells and module components. For the Ref-SHJ design, the expensive metallization layout results in higher module costs.

As a result, only the PVD-SHJ design has lower total module costs of 0.48 USD/$W_p$compared to 0.50 USD/$W_p$ for the Ref-cSi module. The other SHJ designs have cost ranging from 0.51 for the IBC-SHJ design, to 0.56 USD/$W_p$ for the ALD-SHJ design.

For the SHJ designs, the wafer and module contribute about 36–39% and 35–38% of the overall costs, respectively. Cell processing contributes 23–24%, for the designs with an PVD Ag back contact (PVD-, NE- and ALD-SHJ), 27% for the IBC-SHJ module, and 28% for the Ref-SHJ module. For the reference c-Si module, due to lower efficiency and cheaper cell processing, the contribution of wafer and module cost (46% and 37%) is much larger compared to that of cell processing (17%).

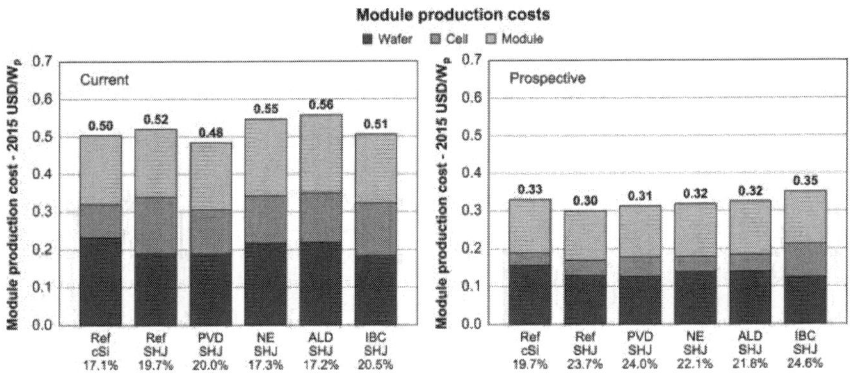


**Figure 9.** Summary of total module cost for the monocrystalline silicon and SHJ designs. Totals are indicated in bold above the bars. A breakdown of cell costs is shown in Fig. 8. For a breakdown of the cost of module elements see Fig. 10. The percentages below the labels indicate module efficiencies.

## 5.2. From current to prospective production costs

In Section 4 we identified several developments and pathways towards reduction of PV production costs. To summarize, these were: (1) Increased cell and module efficiency for all designs (2) Decreased cost of wafers due to higher silicon utilization (thin wafers, less kerf loss) (3) Reduction of silver usage per cell followed by (4) Substitution of silver with copper (5) Replacement of ITO with ZnO, and (6) Slight changes to the module frame.

Fig. 11 shows the effect of these changes on the five module designs studied here. Major cost reductions can be obtained by improved cell efficiency and lower wafer production costs. The increased efficiency of the prospective SHJ designs leads to a cost decrease of 12–16%, while this figure is 9.6% for the reference c-Si design. Higher silicon utilization due to thinner wafers, produced with less silicon kerf losses, allow for a reduction in cost of 9–10% for the SHJ designs (11% for c-Si). The reduction of silver consumption projected by the International Technology Roadmap for Photovoltaics [48], of about 60% in 2025, would allow for a cost reduction of 11% for the silver-heavy Ref-SHJ designs, and 6% for the remaining SHJ designs with screen printed Ag grids. Substitution of silver with copper-based metallization would allow for a further decrease of 4–7%. Silver use reduction (−4%) and substitution (−6%) would not decrease the production costs of the conventional c-Si module as much, as the contribution of metallization cost to overall costs is much lower (see Fig. 8). Replacement of TCO would allow for cost reduction of around 1%, while reduction in the amount of aluminium used for the module frame and a switch to bifacial modules where possible, would allow for a further cost reduction of 3–5% for all designs.

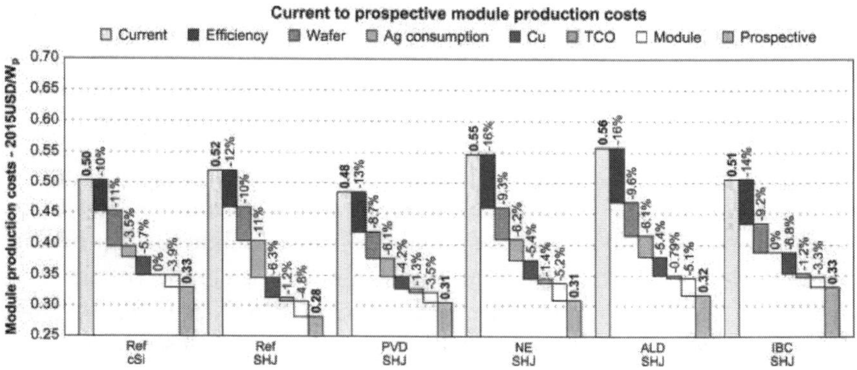


**Figure 11.** Stepped bar chart showing the possible cost reductions achieved by changes in various parameters, from current to prospective designs.

Combined, these developments result in a significant expected drop in cell production costs and total module costs. Cell processing costs drop by 62% for the Ref-cSi design, 73% for the Ref-SHJ design, 67–68% for the ALD-SHJ and NE-SHJ designs, 57% for the PVD-SHJ design, and 36% for the IBC-SHJ design. Total module costs drop by 34% for the c-Si module, 31% for the IBC-SHJ module and, 36–42%, for the other SHJ modules. As Fig. 8 and Fig. 9 show, they would thus allow for cell production cost of 0.19–0.23 USD/$W_p$ and 0.21 USD/$W_p$ for SHJ and c-Si cells respectively, and 0.30–0.35 USD/$W_p$ and 0.33 USD/$W_p$ for SHJ and c-Si modules, respectively.

The IBC module shows the least potential for cost reduction as its' prospective costs are hampered by the requirement for the structured backsheet, the high capex costs, and the high yield and throughput losses incurred by the complex processing sequence.

# 6. SENSITIVITY AND UNCERTAINTY

The cell and module production costs shown above rely heavily on a few major parameters: (1) Uncertainty in cell and thus module efficiency; (2) Historical variation in wafer cost; (3) Variation in silver paste price and uncertainty in silver consumption per cell, and (4) Uncertainty in prices of consumables. We performed a sensitivity and uncertainty analysis, to show the sensitivity of the overall results to changes in these parameters and to assess the effect of uncertainty in these parameters on the overall result. To analyze the sensitivity of the overall results to these parameters, we varied the parameters and plotted the resulting module costs in Fig. 13. For the uncertainty analysis, we performed a Monte Carlo simulation: random samples for each parameter were generated from a log-normal distribution, and the overall module cost was recalculated with these random samples. For instance, we generated 100,000 module efficiencies randomly from a log-normal distribution and recalculated module costs with these 100,000 samples. The resulting ranges of results are represented as box plots inFig. 14 showing the distribution of obtained module costs.

Table 10 shows the parameters we investigated, the ranges over which we varied them, and the standard deviation we used to generate the random samples for the Monte Carlo simulations. The results of the sensitivity and uncertainty analysis are shown inFig. 13 and Fig. 14 respectively.

## 6.1. Cell and module efficiency

As the novel SHJ designs studied here are still under development, accurate measurements of cell efficiency in mass production or even pilot production are not yet available. Furthermore, as shown in Section 5.2 and Fig. 11, a large fraction of prospective cost reductions relies on an assumed increase in cell and module efficiency. Recent development of varying designs show significant variation in SHJ cell efficiency[19].

We assigned an efficiency range to each of the studied designs, assuming that the efficiency of the Ref-cSi design has the smallest uncertainty, being the most established cell design. For the SHJ designs, the range in efficiencies analyzed is larger, especially for the conceptual NE-SHJ and ALD-SHJ designs. The ranges studied are shown inTable 10. These ranges were estimated roughly, based on the deviation of the designs' layout and processing cycle from that of the Ref-SHJ design.

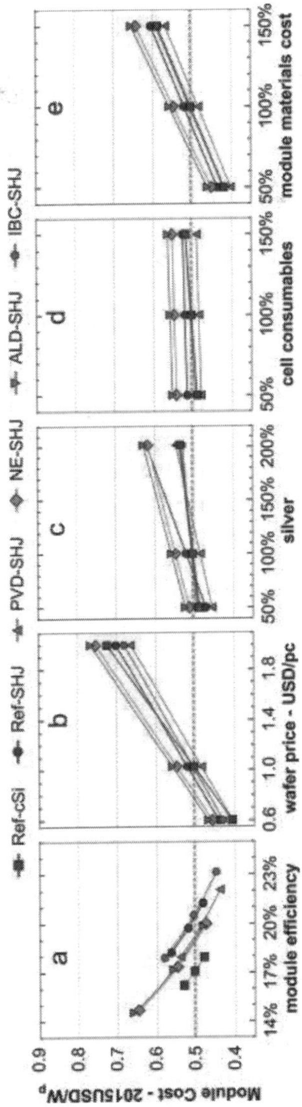


**Figure 13.** Results of the sensitivity analysis. The graphs indicate the change in overall results as a function of change in module efficiency, wafer price, silver price/consumption and consumable price/consumption. The colored markers indicate the minimum, base and maximum results for each of the studied designs. Module efficiency was varied over a range we estimated. Wafer price and silver price were varied over a range reflecting these parameters' historical (5 year) variation. Consumable prices were varied over a range we assumed to account for most variation found on the market. (For interpretation of the references to color in this figure caption, the reader is referred to the web version of this paper.).

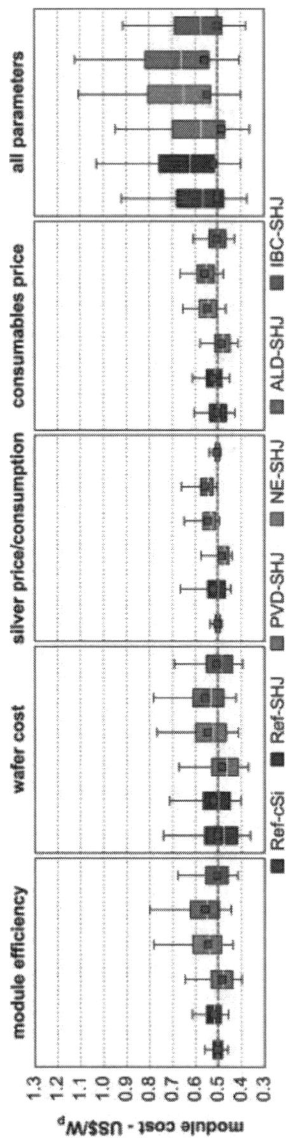


**Figure 14.** Results of the Monte Carlo uncertainty analysis. The box plots indicate the range in overall results obtained when varying efficiency, wafer price, silver price/consumption and consumable price/consumption The red squares and white lines indicate mean and median values respectively, while the boxes indicate 25th and 75th percentiles, and the whiskers indicate minimum and maximum values. The dashed red line indicates the main result for the Ref-cSi design. (For interpretation of the references to color in this figure caption, the reader is referred to the web version of this paper.)

**Table 10.** Overview of parameter ranges studied in the sensitivity analysis (min-max range) and Monte Carlo simulation ($\sigma$).

| Parameter | | Ref c-Si | Ref-SHJ | PVD-SHJ | NE-SHJ | ALD-SHJ | IBC-SHJ |
|---|---|---|---|---|---|---|---|
| Module efficiency | base | 17.1% | 19.7% | 20.0% | 17.3% | 17.2% | 20.5% |
| | min | 16.2% | 18.2% | 17.9% | 14.7% | 14.5% | 17.9% |
| | max | 17.9% | 21.2% | 22.0% | 20.0% | 19.8% | 23.1% |
| | $\sigma$[a] | 0.4% | 0.8% | 1.0% | 1.4% | 1.3% | 1.3% |
| Wafer price | base | 1.03 | 1.03 | 1.03 | 1.03 | 1.03 | 1.03 |
| | min | 0.80 | 0.80 | 0.80 | 0.80 | 0.80 | 0.80 |
| | max | 2.00 | 2.00 | 2.00 | 2.00 | 2.00 | 2.00 |
| | $\sigma$ | 0.128 | 0.255 | 0.255 | 0.255 | 0.255 | 0.255 |
| ag paste price/consumption | base | 100% | 100% | 100% | 100% | 100% | 100% |
| | min | 50% | 50% | 50% | 50% | 50% | 50% |
| | max | 200% | 200% | 200% | 200% | 200% | 200% |
| | $\sigma$ | 0.128 | 0.255 | 0.255 | 0.255 | 0.255 | 0.255 |
| Consumable price/consumption | base | 100% | 100% | 100% | 100% | 100% | 100% |
| | min | 50% | 50% | 50% | 50% | 50% | 50% |
| | max | 150% | 150% | 150% | 150% | 150% | 150% |
| | $\sigma$ | 0.255 | 0.255 | 0.255 | 0.255 | 0.255 | 0.255 |

[a] The sigma (standard deviation) was calculated assuming the respective min–max ranges to be the lower and upper bounds of the 95% confidence interval.

As shown in Fig. 13(a), module cost is quite sensitive to changes in module efficiency. When we varied cell efficiency over the ranges described in Table 10, costs change with a factor of the inverse of the relative efficiency change. The figure clearly shows that the Ref-cSi design has a pronounced processing cost advantage, that is offset only by strong efficiency increases in the PVD-SHJ and IBC-SHJ designs. The other SHJ designs need further efficiency improvements to approach Ref-cSi module costs. At similar efficiency, the SHJ would be more expensive compared to the c-Si module, due to the high cell and for the IBC design also module costs.

## 6.2. Wafer cost

Wafer prices are strongly linked to polysilicon prices, which have shown to be quite volatile. Current (June 2015) spot prices for wafers are about 1.01 USD/wafer, with polysilicon prices at around 15.7 USD/kg [61]. After many silicon producers shutting down the last 1.5 years due to an oversupply, prices were estimated to stabilize and rise slightly to a price of about 20–25 USD/kg [62]. More recently, polysilicon capacity was increased, leading to the possibility of another polysilicon oversupply situation with further price decreases. Polysilicon price is thus quite uncertain. Taking into account the price range of polysilicon for the last 5 years and possible further price decreases, we investigated the effect on the overall results of a wafer price in the range of 0.60 USD/wafer to 2.0 USD/wafer.

Fig. 13(b) shows that all designs are quite sensitive to variation in wafer cost. The effect of changes in wafer cost is similar for all devices, although the effect is more pronounced as the module efficiency decreases, as the wafer cost contributes more to the overall costs at lower efficiencies. This is especially apparent in the results for the Ref-cSi module, and to a lesser degree for the ALD-SHJ and NE-SHJ modules.

## 6.3. Silver paste price and consumption

Another strong contributor to overall cell manufacturing cost is the price of silver paste used for metallization. As shown in Fig. 12, silver prices are quite volatile, with prices ranging from about 500 USD/kg to over 1700 USD/kg over the last five years, with a current (Oct 2015) price of about 520 USD/kg. This silver price variation could account for very large increases in the cost of silver paste, to almost 1500 and 2000 USD/kg for high and low temperature paste, respectively.

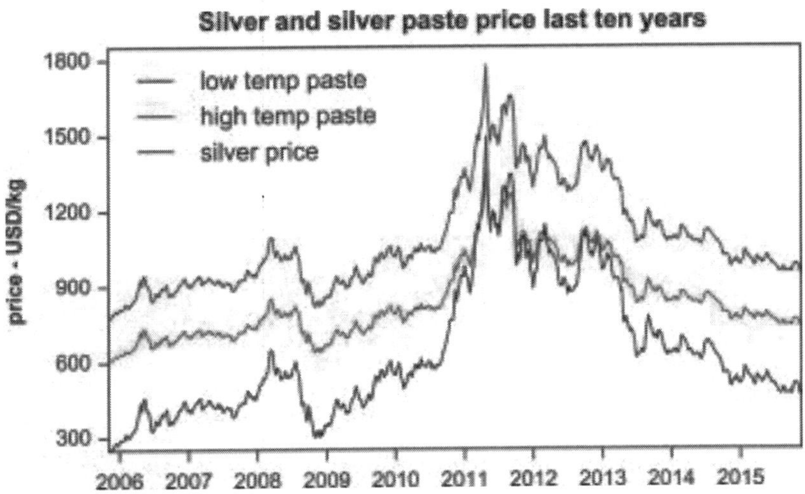


**Figure 12.** Silver price, and estimated high- and low-temperature silver paste price from October 2005 to October 2015. Silver price data: [18] and [63].

Because of the requirement for relatively large amounts of silver in low-temperature paste, the SHJ designs are quite sensitive to changes in paste price or paste consumption per cell, especially the bifacial Ref-SHJ design. A doubling of the paste price would raise the production costs of this designs by up to 20%. The effect is not as strong for the c-Si design, as this design is based on cheaper, high-temperature paste.

## 6.4. Other consumable prices

Price variation can of course occur not only for silver, but also for all other material consumables that are required to produce the solar cells and modules. Therefore, we studied the effect of variation in the prices of cell consumables other than silver paste and wafers, and of module materials on the overall results. As seen in Fig. 13(d), the effect of changes in cell consumable prices is limited, as these consumables only account for around 2–5% of SHJ module prices and 9% of the Ref-cSi module. The results are however very sensitive to changes in the cost of module materials (see Fig. 13(e)), as they contribute around 30–40% of total module cost.

## 6.5. Monte Carlo uncertainty

The results of the Monte Carlo simulation to assess the effect of uncertainty in the data is shown in Fig. 14. This figure shows the distribution of relative cost of the SHJ designs compared to the Ref-cSi design. This figure shows that especially uncertainty in module efficiencies can alter the results (absolute and relative) significantly. Uncertainty in the cost of wafers, silver, and other consumables also has a pronounced effect on module prices (as shown also in Fig. 13(b)–(e)) but changes in these parameters would likely affect all designs similarly and thus not significantly changes the relative results.

# 7. DISCUSSION AND CONCLUSIONS

The main aim of this study is to perform a bottom-up analysis of the relative cost of different silicon heterojunction (SHJ) based PV modules, and to compare current and prospective costs of these modules with those of a conventional monocrystalline silicon module. The aim is to develop a methodology that allows for ex-ante screening of R&D cell concepts. We performed a life cycle costing analysis to analyze in detail the cost structure of the production of PV modules, and found current SHJ modules to be comparable in price compared to conventional monocrystalline silicon modules, but cost penalties incurred by using more expensive materials need high efficiencies to be offset. Thus, especially designs that minimize the use of low-temperature silver paste benefit from the efficiency advantage of heterojunction technology.

The life cycle costing (LCC) method shows its' usefulness in determining the contribution of each process and material input to the overall production cost of different cell and module designs. Our research has confirmed that for all designs investigated wafer and metallization contribute the majority of overall cell production costs, while for complete modules, the module materials are also a very significant cost factor. Reductions in wafer thickness, silver paste usage, or substitution of silver with copper-based metallization allows for significant cost reductions. The strong effect of module efficiency on module price (USD/$W_p$) however shows that changes in cell and module design that have a negative effect on module efficiency will likely result in a relative cost increase. Furthermore, as Fig. 13(a) shows, the efficiency advantage of the studied SHJ designs is the main reason for there lower production costs (in terms of USD/$W_p$). This indicates that: (a) the uncertainty in cell and module efficiency for the SHJ designs results in significant uncertainty of the overall result (see Fig. 13 and Fig. 14), (b) changes in cell processing that decrease cell efficiency are likely to result in cost increases, even if they reduce cell processing costs, (c) accurate cost calculations for SHJ cell concepts should preferably rely on cell efficiency data from gathered from (pilot) production lines.

## 7.1. Wafer cost

To calculate the wafer production cost, we performed an LCC analysis of wafer production from silica to sawn wafers. Our model results for current wafers (1.03 USD/pc) results in slightly higher costs compared to the current wafer spot price (1.01 USD/pc) [61]. This likely is the result of the fact that the life-cycle data on silicon production we used for our calculations is already some years old (from 2007), although we have updated this as much as possible, as progress has been made especially in reducing the energy budget of silicon production. A different issue with our model is that we have used a single cost model for the wafer used, e.g. we have not distinguished between *n*-type and *p*-type wafers. It is sometimes assumed that *n*-type wafers are likely to be more expensive compared to *p*-type wafers because of lower yields, more complex processing and added capital expenses, however this has been disputed [18].

## 7.2. Metallization

In PV industry and research, it is common knowledge that metallization is a main contributor to PV module costs. R&D on alternative metallization schemes is therefore a very active field of research. The amount of silver used per wafer has already dropped significantly [35], a trend which is expected to continue. Our results indicate that a projected reduction in silver use per cell [35] could decrease the costs of the silver-based SHJ modules studied here by about 6% for most designs, but up to 11% for a design with high silver consumption. As conventional c-Si cells use cheaper silver paste, containing less silver, this cost reduction potential is smaller (−4%) for this technology.

Aside from material use reduction, replacement of silver with copper is also being heavily researched, mainly through research on copper-plating. Copper offers similar conductivity, at only a fraction (~2%) of the cost. Furthermore, the supply of copper is less constrained, although not impervious to price volatility. Our results show that silver substitution could decrease module cost by an additional 4–7% for SHJ modules, or 6% for c-Si modules. In our prospective designs, the contribution of metallization is already low, as we assume that paste consumption per wafer to substantially decrease.

The adoption of copper in the manufacturing process is however not without issues: copper rapidly oxidizes during thermal curing, and can diffuse into the silicon cell substrate [35], [53] and [52]. Recent results have however shown that contact curing can be performed with minimal oxidation in inert or low-oxygen curing atmospheres[52] and [50], or in normal atmospheres using a specially developed paste [53]. Furthermore, the ITO layer applied in SHJ solar cells is stated to be an effective barrier to metal diffusion [64]. Especially copper-plated contacts however, in conjunction with a nickel seed layer and silver capping layer, show prospects for high efficiency SHJ solar cells [54].

An issue with our current model for metallization costs is the cost-structure of metallization pastes. There is limited information on the cost of these pastes, but the available data shows that these pastes are much more expensive than pure silver. This indicates that a significant fraction of metallization paste cost is non-silver related (460 and 580 USD/kg for high- and low-temperature paste). For aluminium pastes, this does not seem to be the case, as these pastes are much cheaper (~50 USD/kg). It does however seem to indicate that the cost in front-side metallization paste (especially low-temperature) cannot only be reduced by reducing the silver content or replacing silver with another metal. Alternative forms of metallization, such as SmartWire Connection Technology [65], inkjet or stencil printing or electroplating could possibly offer further metallization cost reductions.

## 7.3. ITO replacement

From a cost perspective, replacement of ITO has very limited benefits, as the amount of indium needed for the thin transparent conductive oxides is very low. Alternative TCO's that are cheaper but less conductive, are likelier to increase the module cost due to decreased cell efficiency. As one of the options, we have also modelled replacement of ITO with ZnO, which offers a very modest cost decrease.

At current price levels (June 2015, 750 USD/kg), ITO contributes about 0.8–1.8% of cell costs and 0.5–1.1% of total module costs. As we have possibly overestimated the cost of ITO target material (see Section 3.2.2), this contribution could be smaller. Thus, from a cost perspective, ITO replacement does not seem to be that important. At a 100% price increase for ITO, the contribution increases slightly to 1.0–2.2% of module costs. The development of alternative TCOs should however still be researched, as the availability of indium could become a concern [59] and [60].

## 7.4. Module design and materials

Our results indicate that module materials already contribute a large fraction of overall module cost, and will likely contribute the majority of module costs as wafer and cell production become cheaper. Cost reduction possibilities in module design are likely limited, as the production of the materials used is already very mature, and there is thus limited possibility for technological learning. For current modules, main costs are glass, frame and backsheet. Prospective glass-glass modules can be somewhat cheaper compared to modules with an aluminium frame, because of the replacement of the backsheet with a sheet of glass. The cost of the glass on the backside of the prospective modules is likely overestimated, as lower quality glass can be applied here compared to the front-side. This could significantly lower the price of the backside of the module. Furthermore, the costs calculated here are based on monofacial efficiencies, and thus do not reflect the additional energy yield than can be obtained from bifacial PV modules.

## 7.5. Concluding remarks

Our results show that silicon heterojunction (SHJ) technology offers the potential for cost reductions in PV manufacturing compared to conventional crystalline silicon solar cells and modules, especially for prospective PV modules. Heterojunction module production costs were found to range from 0.48 to 0.56 USD/$W_p$, compared to 0.50 USD/$W_p$ for a conventional monocrystalline silicon module. Heterojunction modules incur a strong cost penalty because of the requirement for low-temperature silver paste. High-efficiency SHJ designs that minimize the use of this paste can be competitive with standard crystalline silicon PV modules. As the designs studied are conceptual, the validity of the results will strongly benefit from more accurate determination of the efficiency of SHJ cells and modules from large-scale production. Variation of material prices can also have a large influence on the cost of SHJ modules, but these variations will likely affect all designs similarly and thus not alter the relative results.

Our results confirm that the choice of metallization has a substantial impact on the cost of SHJ modules compared to conventional crystalline silicon modules. This is mainly due to the high cost of low-temperature paste needed for SHJ cell processing, and the increased amount of paste required due to its lower as-cured conductivity. This results in higher cell costs for SHJ designs (USD/cell), that is offset partly by the high efficiency of heterojunction technology.

Our prospective analysis indicates that a reduction in paste consumption and replacement of silver with copper paste could alleviate this problem, e.g., the cost-advantage of SHJ cells becomes less dependent on the efficiency advantage. However, these results are based on the assumption that progress in reduction of silver consumption is also applicable to SHJ cells. A more promising approach seems to be the replacement of screen printed silver grids with copper-electroplated contact, minimizing the amount of silver used in SHJ cells. For all studied prospective modules, the module materials contribute significantly more to the overall module cost than they do now.

Interdigitated back-contacted (IBC) design are often mentioned as a promising low-cost (future) module technology. Our results however show that for now, the requirements for this technology during cell processing and module production cannot be completely offset by the large efficiency advantage obtained through the improved cell structure.

As the prospective cost reductions that we have modelled are very dependent on reductions and eventual elimination of silver use in SHJ cells, further research should keep addressing these issues, by establishing a metallization process based on copper that is compatible with industrial scale production without adding complexity to the production process. Furthermore, increasing attention should go towards reducing the cost of module elements.

## ACKNOWLEDGMENTS

This work was performed within "FLASH", a *Perspectief* programme funded by Technology Foundation STW (www.stw.nl/en) under project number 12172. The authors would like to thank the reviewers for their very detailed and helpful comments, as they feel these comments strongly contributed to the quality of the paper.

## REFERENCES

1. W.G. van Sark, P. Muizebelt, J. Cace, A. de Vries, P. de Rijk, Price development of photovoltaic modules, inverters, and systems in the Netherlands in 2012, Renew. Energy 71 (2014) 18–22,

2. C. Breyer, A. Gerlach, Global overview on grid-parity, Prog. Photovolt.: Res. Appl. 21 (1) (2013) 121–136,

3. 3] N.S. Lewis, Toward cost-effective solar energy use, Sci. New York, N.Y. 315 (5813) (2007) 798–801,

4. M. Bazilian, I. Onyeji, M. Liebreich, I. MacGill, J. Chase, J. Shah, D. Gielen, D. Arent, D. Landfear, S. Zhengrong, Re-considering the economics of photovoltaic power, Renew. Energy 53 (2013) 329–338,

5. C. Budischak, D. Sewell, H. Thomson, L. Mach, D.E. Veron, W. Kempton, Costminimized combinations of wind power, solar power and electrochemical storage powering, the grid up to 99.9% of the time, J. Power Sources 225 (2013) 60–74,

6. S. Reichelstein, M. Yorston, The prospects for cost competitive solar PV power, Energy Policy 55 (2013) 117–127,

7. K.D. Swift, A comparison of the cost and financial returns for solar photovoltaic systems installed by businesses in different locations across the United States, Renew. Energy 57 (2013) 137–143,

8. PV Insights, PVinsights: Solar Photovoltaic (Polysilicon wafer cell and panel) Prices and Research Reports, 2014.

9. D.D. Hsu, P. ODonoughue, V. Fthenakis, G.A. Heath, H.C. Kim, P. Sawyer, J.- K. Choi, D.E. Turney, Life cycle greenhouse gas emissions of crystalline silicon photovoltaic electricity generation, J. Ind. Ecol. 16 (2012) S122–S135,

10. H.C. Kim, V. Fthenakis, J.-K. Choi, D.E. Turney, Life cycle greenhouse gas emissions of thin-film photovoltaic electricity generation, J. Ind. Ecol. 16 (00) (2012) S110–S121,

11. N. Reich, E. Alsema, W. van Sark, W. Turkenburg, W. Sinke, Greenhouse gas emissions associated with photovoltaic electricity from crystalline silicon modules under various energy supply options, Prog. Photovolt.: Res. Appl. 19 (5) (2011) 603–613,

12. V. Fthenakis, H. Kim, Photovoltaics: life-cycle analyses, Solar Energy 85 (8) (2011) 1609–1628,

13. Louwen, W.G.J.H.M. van Sark, R.E.I. Schropp, W.C. Turkenburg, A.P.C. Faaij, Life-cycle greenhouse gas emissions and energy payback time of current and prospective silicon heterojunction solar cell designs, Prog. Photovolt.: Res. Appl. 23 (10) (2015) 1406–1428,

14. International Energy Agency, World Energy Outlook 2014, International Energy Agency, Paris, France, 2014.

15. H. Wirth, Recent facts about photovoltaics in Germany, Technical Report, Fraunhofer ISE, Freiburg, Germany, 2015.

16. K. Blok, Introduction to Energy Analysis, Techne Press, Amsterdam, The Netherlands, 2007.

17. OpenEI, Transparent Cost Database – Transparent) Cost Database, URL ⟨http:// en.openei.org/apps/TCDB/⟩.

18. Goodrich, P. Hacke, Q. Wang, B. Sopori, R. Margolis, T.L. James, M. Woodhouse, A wafer-based monocrystalline silicon photovoltaics road map: utilizing known technology improvement opportunities for further reductions in manufacturing costs, Sol. Energy Mater. Sol. Cells 114 (2013) 110–135.

19. W. van Sark, L. Korte, F. Roca, Physics and Technology of AmorphousCrystalline Heterostructure Silicon Solar Cells, Engineering Materials, Springer Berlin Heidelberg, Berlin, Heidelberg, 2012

20. S. De Wolf, A. Descoeudres, Z.C. Holman, C. Ballif, High-efficiency silicon heterojunction solar cells: a review, Green 2 (1) (2012) 7–24,

21. ISO, ISO 14044:2006—Life cycle assessment—Requirements and guidelines, Technical Report, International Organization for Standardization (ISO), Geneva, Switzerland. URL ⟨http://www.iso.org/iso/catalogue_detail?csnumber¼38498⟩, 2006.

22. ISO, ISO 14040:2006—Environmental management—Life Cycle Assessment— Principles and Framework, Technical Report, International Organization for Standardization (ISO), Geneva, Switzerland. URL ⟨http://www.iso.org/iso/cata logue_detail.htm?csnumber¼37456⟩, 2006.

23. V.M. Fthenakis, R. Frischknecht, M. Raugei, H.-C. Kim, E.A. Alsema, M. Held, M. de Wild-Scholten, Methodology guidelines on life cycle assessment of photovoltaic electricity, 2nd edition, Technical Report, IEA PVPS Task 12, International Energy Agency Photovoltaic Power Systems Programme, Paris, France, 2011.

24. A.L. Roes, E. Marsili, E. Nieuwlaar, M.K. Patel, Environmental and cost assessment of a polypropylene nanocomposite, J. Polym. Environ. 15 (3) (2007) 212–226,

25. G. Rebitzer, S. Seuring, Methodology and application of life cycle costing, Int. J. Life Cycle Assess. 8 (2) (2003) 110–111,

26. D. Hunkeler, G. Rebitzer, Life cycle costing—paving the road to sustainable development? Int. J. Life Cycle Assess. 8 (2) (2003) 109–110,

27. PHOTON Europe GmbH, Market survey on wet etching equipment, Photon Int., December 2011, 166–177.

28. PHOTON Europe GmbH, Market survey on deposition equipment for transparent conductive oxide layers, Photon Int., March 2011, 160–175.

29. PHOTON Europe GmbH, Market survey on gas abatement systems for thin- film applications, Photon Int., November 2010, 136–141.

30. PHOTON Europe GmbH, Market survey on silicon thin-film absorber deposition equipment, Photon Int., November 2011, 166–181.

31. PHOTON Europe GmbH, SiNx deposition systems survey, Photon Int., March 2014, 50–61.

32. USGS, 2010 Minerals yearbook silica, Technical Report, U.S. Geological Survey.

33. N. Jungbluth, M. Stucki, R. Frischknecht, Photovoltaics, Sachbilanzen von Energiesystemen: Grundlagen fur den okologischen Vergleich von Energiesystemen und den Einbezug von Energiesystemen in Okobilanzen fur die Schweiz. ecoinvent report No. 6-XII (6-XII).

34. M. Woodhouse, R. Fu, T. James, K. Horowitz, B. Gallagher, D. Feldman, D. Chung, R. Margolis, Current Results from NRELs Wafer-Based Monocrystalline Silicon PV Module Manufacturing Cost Models Current Results from NRELs Wafer-Based Monocrystalline Silicon PV Module Manufacturing Cost Models, in: NREL crystalline silicon workshop, July 2014, Breckenridge, Colorado, USA, 2014.

35. ITRPV working group, International Technology Roadmap for Photovoltaic (ITRPV)—2013 Results, Technical Report, March, 2014.

36. J. Tyler, Manufacturers will slowly switch to diamond wire, solar industry magazine.

37. G. Roters, J. Krause, S. Leu, A. Richter, B. Strahm, Heterojunction technology the solar cell of the future, 2014.

38. Z.C. Holman, A. Descoeudres, S. De Wolf, C. Ballif, Record infrared internal quantum efficiency in silicon heterojunction solar cells with dielectric/metal rear reflectors, IEEE J. Photovolt. 3 (4) (2013) 1243–1249,

39. PHOTON Europe GmbH, Aluminum oxide deposition systems survey, Photon Int., March 2014, 62–81.

40. M. Scherff, Novel method for preparation of interdigitated back contacted a-Si: H/cSi heterojunction solar cells, in: H. Ossenbrink, A. Jäger-Waldau, P. Helm (Eds.), Proceedings of the 26th European Photovoltaic Solar Energy Conference and Exhibition, WIP-Renewable Energies, Munich, Germany, Hamburg, Germany, 2011, pp. 2125–2129.

41. M. Späth, D. Veldman, N. Dekker, I. Bennett, W. Eerenstein, P. de Jong, P. Ribeyron, S. Harrison, D. Muñoz, Heterojunction module technology, In: P. Helm, A. Jäger-Waldau, H. Ossenbrink (Eds.), Proceedings of the 26th European Photovoltaic Solar Energy Conference and Exhibition, WIP-Renewable Energies, Munich, Germany, Hamburg, Germany, September 2011. [42] Y. Zemen, S. Schulz, H. Trommler, S. Buschhorn, W. Bauhofer, K. Schulte, Comparison of new conductive adhesives based on silver and carbon nanotubes for solar cells interconnection, Sol. Energy Mater. Sol. Cells 109 (2013) 155–159,

42. PHOTON Europe GmbH, Cell testers & sorters survey, Photon Int., October 2014, 66–80.

43. PHOTON Europe GmbH, Survey lay up machines, Photon Int., October 2012, 102–117.

44. PHOTON Europe GmbH, Survey laminators, Photon Int., August 2012, 100– 103; 228–275.

45. PHOTON Europe GmbH, Cell interconnection equipment survey, Photon Int., September 2014, 52–70.

46. C.J.J. Tool, E.J. Kossen, I.J. Bennett, M.J.H. Kloos, W. Eerenstein, Improved effi- ciency and cost of ownership for p-type MWT cell and module technology.

47. ITRPV working group, International technology roadmap for photovoltaic (ITRPV) 2014 results, Technical Report, SEMI, April 2015.

48. M.A. Green, Ag requirements for silicon wafer-based solar cells, Prog. Photovolt.: Res. Appl. 19 (8) (2011) 911–916,

49. D. Wood, N. Powell, A. Zambova, B. Chislea, P. Chevalier, C. Boulord, A. Beucher, N. Zeghers, G. Beaucarne, I. Kuzma-filipek, R. Russell, F.

Duerinckx, J. Szlufcik, W. Deng, Z. Feng, P.J. Verlinden, Passivated busbars from screen-printed low temperature copper paste, 2014.

50. Kohn, M. Eizenberg, Diffusion barriers for ultra-large-scale integrated copper metallization, in: Advanced Nanoscale ULSI Interconnects: Fundamentals and Applications, Springer New York, New York, NY, 2009, pp. 93–120.

51. Clement, H. Bell, F. Vogg, L. Rebenklau, P. Gierth, U. Partsch, Inert drying system for copper paste application in PV, Energy Procedia 38 (2013) 423–429,

52. M. Yoshida, H. Tokuhisa, U. Itoh, T. Kamata, I. Sumita, S. Sekine, Novel lowtemperature-sintering type Cu-alloy pastes for silicon solar cells, Energy Procedia 21 (October (2011)) (2012) 66–74,

53. L. Tous, R. Russell, J. Das, R. Labie, M. Ngamo, J. Horzel, H. Philipsen, J. Sniekers, K. Vandermissen, L. van den Brekel, T. Janssens, M. Aleman, D. van Dorp, J. Poortmans, R. Mertens, Large area copper plated silicon solar cell exceeding 19.5% efficiency, Energy Procedia 21 (October (2011)) (2012) 58–65,

54. PHOTON Europe GmbH, Plating equipment survey, Photon Int., July 2013, 118–125.

55. B. Andersson, Materials availability for large-scale thin-film photovoltaics, Prog. Photovolt.: Res. Appl. 8 (May (1999)) (2000) 61–76,

56. L. Grandell, A. Thorenz, Silver supply risk analysis for the solar sector, Renew. Energy 69 (2014) 157–165,

57. Zuser, H. Rechberger, Considerations of resource availability in technology development strategies: the case study of photovoltaics, Resour. Conserv. Recycl. 56 (1) (2011) 56–65,

58. Candelise, J.F. Speirs, R.J. Gross, Materials availability for thin film (TF) PV technologies development: a real concern?, Renew. Sustain. Energy Rev. 15 (9) (2011) 4972–4981,

59. M.Z. Jacobson, M.A. Delucchi, Providing all global energy with wind, water, and solar power, Part I: technologies, energy resources, quantities and areas of infrastructure, and materials, Energy Policy 39 (3) (2011) 1154–1169,

60. PV Insights, PVinsights: Solar Photovoltaic (Polysilicon wafer cell and panel) Prices and Research Reports. URL ⟨www.pvinsights.com⟩, 2015.

61. PV-Magazine, Polysilicon industry to resume growth in 2013. URL ⟨http:// www.pv-magazine.com/news/details/beitrag/polysilicon-industry-to-resumegrowth-in-2013_100009775/#axzz2V9zGHOLF⟩, 2013.

62. London Bullion Market Association, LBMA—pricing and statistics. URL ⟨http:// www.lbma.org.uk/pricing-and-statistics⟩, 2015.

63. F. Roca, G. Sinno, G. Di Francia, P. Prosini, G. Fameli, P. Grillo, A. Citarella, F. Pascarella, D. della Sala, Process development of amorphous silicon/crystalline silicon solar cells, Sol. Energy Mater. Sol. Cells 48 (1–4) (1997) 15–24,

64. Meyer Burger, SWCT—Meyer Burger technology AG. URL ⟨http://www.meyer burger.com/en/products-systems/technologies/ photovoltaic/swct/⟩, 2015.

65. PHOTON Europe GmbH, Diffusion furnaces survey, Photon Int., August 2013, 68–78

66. PHOTON Europe GmbH, Screen printers and printing lines survey, Photon Int., July 2014, 68–79; 102–111.

67. SunSirs, SunSirs, China Commodity Data Group, provides Bulk Commodity Prices & Commodity Markets & Commodity Information & Commodity News. URL ⟨www.sunsirs.com⟩, 2014.

68. N.N.; Tempress, Private communications, 2013.

69. Sigma-Aldrich, Boron trifluoride 99.5%—Sigma-Aldrich. URL ⟨http://www.sig maaldrich.com/catalog/product/aldrich/295027? lang¼en&region¼US⟩, 2014.

70. Sigma-Aldrich, Zinc oxide — Sigma-Aldrich. URL https://www. sigmaaldrich. com/catalog/substance/zincoxide8139131 413211?lang ¼en&region¼US, 2014.

71. U.S. Energy Information Administration, U.S. Weekly Heating Oil and Propane Prices (October–March), 2014.

72. Sigma-Aldrich, Trimethylaluminum 97% — Sigma-Aldrich. URL http:// www. sigmaaldrich.com/catalog/product/aldrich/257222? lang¼ en& region¼US, 2014.

# Chapter 7

# Minority carrier lifetime in silicon photovoltaics: The effect of oxygen precipitation

**J.D. Murphy [a,n] , R.E. McGuire [b] , K. Bothe [c] , V.V. Voronkov [d] , R.J. Falster [d]**

[a] University of Warwick, WMG, International Manufacturing Centre, Coventry, CV4 7AL, UK

[b] University of Oxford, Department of Materials, Parks Road, Oxford, OX1 3PH, UK

[c] Institut für Solarenergieforschung Hameln/Emmerthal, Am Ohrberg 1, 31860 Emmerthal, Germany

[d] SunEdison, viale Gherzi 31, 28100 Novara, Italy

## ABSTRACT

Single-crystal Czochralski silicon used for photovoltaics is typically supersaturated with interstitial oxygen at temperatures just below the melting point. Oxide precipitates therefore can form during ingot cooling and cell processing, and nucleation sites are typically vacancy-rich regions. Oxygen precipitation gives rise to recombination centres, which can reduce cell efficiencies by as much as 4% (absolute). We have studied the recombination behaviour in p-type and n-type monocrystalline silicon with a range of doping levels intentionally processed to contain oxide precipitates with a range of densities, sizes and morphologies. We analyse injection-dependent minority carrier lifetime measurements to give a full parameterisation of the recombination activity in terms of Shockley–Read–Hall statistics. We intentionally contaminate specimens with iron, and show recombination activity arises from iron segregated to oxide precipitates and surrounding defects. We find that phosphorus diffusion gettering reduces the recombination activity of the precipitates to some extent. We also find that bulk iron is preferentially gettered to the phosphorus diffused layer rather than to oxide precipitates.

## KEYWORDS

Silicon; Photovoltaic; Oxide precipitate; Lifetime; Recombination

## 1. INTRODUCTION

At present the majority of solar cells are made from bulk crystalline silicon. Minority carrier lifetime is the main parameter used to assess the quality of wafers from which cells are produced. For a given generation rate, the minority carrier lifetime is largely determined by recombination processes. Some recombination is intrinsic (band-to-band and Auger), while other is determined by defects in the bulk or at surfaces. It is necessary to understand which defects are typically present in solar wafers before processing, and what effect processing has on those defects. Moreover, it is important to understand the mechanism by which the relevant defects give rise to recombination, as well as to quantify their recombination activity.

Monocrystalline Czochralski silicon (Cz-Si) typically contains ~$10^{18}$ $cm^{-3}$ of interstitial oxygen, which is mainly incorporated from the silica crucible which contains the melt. This level of oxygen is supersaturated below ~1200 °C, so the equilibrium state is reached by the formation of $SiO_2$ particles (oxide precipitates) [1]. The morphology of such particles changes as they grow, from unstrained particles initially, to strained precipitates, which are eventually surrounded by dislocations and sometimes stacking faults [2] and [3]. The first stage in precipitation is nucleation, and the rate of this is strongly enhanced by the presence of crystal defects. In modern Cz-Si used for integrated circuits the grown-in defect concentration is insufficient for oxygen precipitation to occur unintentionally. However, even in the highest quality Cz-Si, oxide precipitates can nucleate upon prolonged annealing at 650 °C to 850 °C [1] and [4]. In silicon for microelectronics thermal processes are often used to force oxygen precipitation to provide gettering centres for harmful metallic contaminants [5] and [6]. For photovoltaics, the Cz-Si wafers used are often of lower crystal quality, and several studies have found concentric rings of oxide precipitates in wafers or cells after growth or processing [7], [8], [9] and [10]. It is also noted that oxygen precipitation occurs at dislocations in multicrystalline silicon (mc-Si) during ingot cooling [11], [12] and [13].

The undesirable precipitation of oxygen in Cz-Si is not a new problem, having been widely studied (and essentially solved) in Cz-Si for integrated circuits. Oxide precipitates are known to form in vacancy-rich regions [14]. The formation of such regions can be essentially eliminated by carefully controlling the so-called $v/G$ criterion (where $v$ is the crystal growth rate and $G$ is the near-interface temperature gradient) [15]. Although it is possible to reduce intrinsic point defect concentrations to levels which are essentially negligible, doing so requires slow growth rates and these are not always compatible with the commercial constraints of the silicon photovoltaics industry. Thus, although the problem can in principle be eliminated, the fact is many commercial Cz-Si solar wafers do contain vacancy-rich regions in which oxide precipitates form [7], [8] and [9].

Oxide precipitates have been linked to a substantial detrimental impact on conversion efficiencies in silicon solar cells [7] and [16]. A study by Haunschild et al. associated oxide precipitates with a 4% (absolute) efficiency reduction [7]. Interestingly, they found that the recombination activity to be strongly affected by a 10 s anneal at 800 °C. This, and earlier studies on iron-contaminated samples [17] and [18], suggests that impurities might play a role in the recombination mechanism.

The aim of this paper is to answer some open questions regarding the effects of oxygen precipitation in silicon photovoltaics. These include:

- Can the impact of oxide precipitates on minority carrier lifetime be systematically quantified?
- What is the mechanism of recombination at oxide precipitates? Do impurities play a role?
- What happens to the recombination activity of oxide precipitates after phosphorus diffusion gettering used in solar cell processing? Which is

the more effective gettering system for bulk iron: oxide precipitates or phosphorus diffusion?

This paper builds upon our previously published work in this area [19], [20], [21],[22] and [23] by assimilating the key ideas in one article, adding new experimental data, and focussing specifically on photovoltaics. After presenting a general methodology for parameterising injection-dependent minority carrier lifetime in semiconductors, we apply this methodology to determine the recombination parameters of oxide precipitates in silicon.

# 2. PARAMETERISATION OF CARRIER LIFETIMES

## 2.1. Linear formulation of Shockley–Read–Hall statistics

Shockley–Read–Hall (SRH) statistics [24] and [25] are frequently used to quantify the bulk minority carrier lifetime in semiconductor materials. SRH statistics enable the recombination activity of states associated with point-like defects to be quantified by using just three parameters: the energy position of the defect in the bandgap (*ET*), its capture coefficient for electrons ($\alpha n$) and its capture coefficient for holes ($\alpha p$). (Alternatively capture cross-sections for electrons ($\sigma n$) and holes ($\sigma p$) can be used instead of capture coefficients. The capture coefficient is the product of the capture cross-section and the thermal velocity). Capture coefficients (or cross-sections) are temperature-dependent empirically-determined parameters which quantify the propensity of the states to capture carriers. In some circumstances the use of simple SRH statistics is an oversimplification (see for example [26]), but for the most part the SRH approach is invaluable.

The most commonly used formation of SRH statistics describes the minority carrier lifetime in terms of the excess concentration of minority carriers ($\Delta n$ for electrons; $\Delta p$ for holes) (see Ref. [27] for example). In our work we use a different form which we have derived in an earlier paper [21]. We express the minority carrier lifetime as a *linear*function of the ratio of the total carrier concentrations. In p-type material, for example, the electron lifetime ($\tau n$) is expressed as a linear function of the ratio of the total electron concentration ($n=n_0+\Delta n$) to the total hole concentration ($p=p_0+\Delta p$) according to:

$$\tau_n = \frac{1}{\alpha_n N}\left[1 + \frac{Qn_1}{p_0} + \frac{p_1}{p_0} + X\left(Q - \frac{Qn_1}{p_0} - \frac{p_1}{p_0}\right)\right] \quad (1)$$

where X=(n/p)=(1/$p_0$+n), where $p_0$ is the equilibrium hole concentration, $n_0$ is the equilibrium electron concentration, Q=($\alpha_n/\alpha_p$)=($\sigma_n/\sigma_p$), and $N$ is the concentration of the defect. The so-called SRH densities for electrons ($n_1$) and holes ($p_1$) are given by:

$$n_1 = N_C \exp\left(-\frac{E_C - E_T}{kT}\right) \quad (2)$$

$$p_1 = N_V \exp\left(-\frac{E_T - E_V}{kT}\right) \quad (3)$$

where *EC* and *EV* are the energies of the conduction band and valence band edge, respectively, and *NC* and *NV* are the densities of states in the conduction band and valence band, respectively. An equation analogous to Eq. (1) for the hole lifetime ($\tau p$) in n-type material can be derived as [21]:

$$\tau_p = \frac{1}{\alpha_p N}\left[1 + \frac{n_1}{n_0} + \frac{p_1}{n_0 Q} + Y\left(\frac{1}{Q} - \frac{n_1}{n_0} - \frac{p_1}{Qn_0}\right)\right] \quad (4)$$

where Y=(p/n)=(p/$n_0$+p).

## 2.2. Extracting defect parameters from injection-dependent lifetime data

The linear formulation of SRH statistics provides an elegant route to parameterise recombination due to specific defects. The key feature of Eqs. (1) and (4) is that all the injection-dependence of the minority carrier lifetime is consumed into *X* or *Y*. All other terms in the Equations depend upon the properties of the defect, material or temperature. In the p-type case, information on the key SRH parameters can be extracted by taking the derivative of Eq. (1) with respect to *X* and dividing this by the high injection limit of lifetime ($\tau n$ as $X \rightarrow 1$), which gives:

$$\frac{d\tau_n}{dX} / \tau_{nX \rightarrow 1} = \frac{Q}{1+Q} - \frac{1}{p_0}\left(\frac{Qn_1 + p_1}{1+Q}\right) \quad (5)$$

By studying material with different values of $p_0$, Eq. (5) can be used to deduce values of $Q$ and $Qn_1+p_1$, and an example to show this is given in Section 2.3 below. Information can also be gained from inspecting the low injection limit of Eq. (1) ($\tau n$ as $X\to0$), which gives:

$$\tau_{nX\to0} = \frac{1}{\alpha_n N}\left[1+\frac{1}{p_0}(Qn_1+p_1)\right] \tag{6}$$

If $Qn_1+p_1$ is known it is thus possible to use Eq. (6) to extract $\alpha nN$, which is useful as, for constant $\alpha n$, it is proportional to the state density.

It is important to note that from single temperature injection-dependent lifetime measurements alone it is not possible to extract unique values for $\alpha n$, $\alpha p$, or $ET$. Such measurements on samples with different doping levels enable the deduction of $Q=(\alpha_n/\alpha_p)$ and $Qn_1+p_1$ (which depends upon $ET$ via Eqs. (2) and (3)). To isolate separate values of $\alpha n$ and $\alpha p$ it is necessary to determine the state density, $N$, by another technique. Temperature-dependent lifetime measurements can be used to determine whether $Qn_1$or $p_1$ dominates, and hence can allow $ET$ to be uniquely found.

In principle the linear SRH formulation is valid at all injection levels. However, it is important to note that care must be taken when using the approach over a wide range of injection levels. Under typical conditions, $X$=0.01 corresponds to $\Delta n\approx0.011p_0$. Thus processes which manifest themselves at low injection, such as trapping [28] and surface effects [29], can occur very close to the $X\to0$ limit.

## 2.3. Example of FeB in silicon

We demonstrate the merits of this linear SRH formulation by considering the well-established example of FeB pairs in silicon. In Fig. 1(a) we have simulated the injection-dependent lifetime due to (only) this defect. We note that different values of the SRH parameters for FeB exist in the literature [30], [31] and [32]. For our example here we use the parameters of Rein and Glunz [30], which are $ET=EC-0.26$ eV, $\sigma n=2.5\times10^{-15}$ cm$^2$,$\sigma p=5.5\times10^{-15}$ cm$^2$, and we have used $2\times10^7$ cm s$^{-1}$ for the thermal velocity, 25 °C for the temperature and a defect concentration of $1\times10^{12}$ cm$^{-3}$. Values of $NC$ and $NV$ were taken from Ref. [33]. The first feature to note in Fig. 1(a) is that the lifetime plotted against $X$ is linear (Eq. (1)). Second, the lifetime at high injection (as $X\to1$) tends to the so called 'ambipolar' lifetime given by Eq. (1) as $(1/\alpha_n+1/\alpha_p)/N$, which is independent of the doping level. Third, at low injection (as $X\rightarrow 0$), the lifetime is dependent upon the doping level. It would also be dependent upon the defect's energy level.

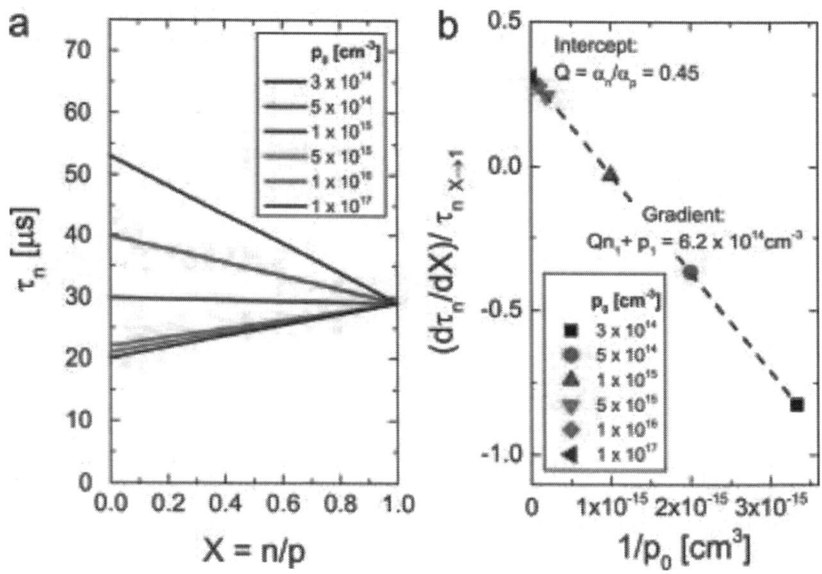


**Figure 1.** (a) Simulated data for the injection-dependence of electron lifetime due to the FeB pair in p-type silicon with different doping levels. The simulations use the SRH parameters of Rein and Glunz [30], with a FeB pair concentration of $10^{12}$ cm$^{-3}$, a temperature of 25 °C and a thermal velocity of $2\times10^7$ cm s$^{-1}$. (b) A plot using Eq. (5) to deduce the values of $Q$ and $Qn_1+p_1$ from the simulated data in (a).

Fig. 1(b) is a plot of the gradient of the data plotted in Fig. 1(a) normalised by the high injection lifetime, plotted against the reciprocal of doping in accordance with Eq. (5). The y-intercept of depends only upon $Q$, whereas the gradient depends only upon $Q$ and$Qn_1+p_1$. Unsurprisingly the plot in Fig. 1(b) gives the characteristic $Q$ value for the FeB pair and the $Qn_1+p_1$ value consistent with that of the FeB pair. Other information (such as temperature-dependent measurements) is needed to isolate whether the $Qn_1$ or the $p_1$term is dominant. The same methodology is drawn upon in this paper to quantify recombination at oxide precipitates in silicon, for which the parameters are not well established.

# 3. EXPERIMENTAL METHODS

## 3.1. Sample production

A set of samples with different oxide precipitate densities, sizes and morphologies were produced from high purity integrated circuit wafers with different doping levels, types and interstitial oxygen concentrations. A four-stage precipitation treatment, as described in detail in Ref. [19], is applied to the samples under test. P-type samples were doped with $3.9\times10^{14}$ cm$^{-3}$ to $8.2\times10^{15}$ cm$^{-3}$ of boron. N-type samples were doped with $5\times10^{13}$ cm$^{-3}$to $1.0\times10^{15}$ cm$^{-3}$ of phosphorus. Strained oxide precipitate densities ($N_{strained}$) were measured by Schimmel etching. Precipitate densities ranged from $3\times10^{6}$ cm$^{-3}$ to $7\times10^{10}$ cm$^{-3}$. Some of the p-type samples have been characterised by transmission electron microscopy (TEM) and the results are discussed in detail elsewhere [3] and [19]. The TEM study performed on a subset of the samples measured enabled the identification of samples in which dislocations, and in some cases stacking faults, were found to surround the oxide precipitates.

Some of the data presented in this paper were obtained using samples in which substantial denuded zones free of oxide precipitates existed near the surfaces. This typically had a depth of ~15 μm to ~35 μm on each side. The denuded zones were removed prior to processing using a planar chemical etch (HF (40% aqueous by volume), $HNO_3$ (69% aqueous by volume) and $CH_3COOH$ (100%) mixed in the volume ratio 8:75:17). Each sample was etched for 40 min four times, so that between 40 μm and 70 μm of material was removed from each side.

Great care was taken to avoid contamination of the samples by metallic impurities during the oxygen precipitation process. However, some samples were intentionally subsequently contaminated with iron, using the same procedure as in Refs. [23],[34] and [35]. This involved rubbing the back-side of the sample with iron pieces (99.95% purity from Testbourne Limited, UK). Samples were then annealed in air in a pre-heated furnace at temperatures up to 798 °C for times chosen to ensure complete iron diffusion through the sample. Although our intention was to contaminate samples only with iron, the possibility that other transition metal impurities have entered the samples cannot be completely ruled out. Cooling was rapid, with the samples removed from the furnace at temperature and placed on a heat sink. It is estimated that samples were cooled to below ~100 °C in <10 s. Samples which were not intentionally contaminated with iron are referred to as "uncontaminated" samples in this paper, although it is possible that the low levels of impurity contamination in such samples have significant effects, as discussed later.

## 3.2. Surface passivation and measurement of lifetime

Samples were cleaved into 3.5 cm by 3.5 cm or 5 cm by 5 cm pieces, which were RCA cleaned. Silicon nitride was then deposited on both surfaces by plasma enhanced chemical vapour deposition (PECVD). Two different PECVD processes were used for the results presented in this paper. The data presented in Fig. 3, Fig. 4 and Fig. 5 were from samples passivated by remote plasma PECVD at ISFH, which has previously been shown to give a surface recombination velocity below 10 cm $s^{-1}$[36]. The data presented in Fig. 6, Fig. 7 and Fig. 8 were from samples passivated using a direct plasma Oxford Instruments Plasmalab 80 Plus PECVD system at the University of Oxford. The surface recombination velocity associated with this latter scheme has not been studied in detail, but is estimated to be of order 100 cm $s^{-1}$. The lower quality scheme was used for samples with lower bulk lifetimes. The lifetime data for such samples could be fitted with the same recombination parameters extracted from the samples for which the better surface passivation scheme was used. It is possible that both PECVD processes introduces hydrogen into the bulk, which may passivate certain defects [37].

Minority carrier lifetime was measured using transient or quasi-steady-state photoconductance [38] methods, with a Sinton WCT-120 lifetime tester. The injection level range studied varied with the lifetime of the sample, but was usually in the range $10^{13}$ $cm^{-3}$ to $10^{16}$ $cm^{-3}$. It is our aim to determine the absolute lifetime associated with oxygen precipitation, so care was taken to prevent or factor out well-understood recombination processes. Boron–oxygen defects were eliminated by storing the samples in the dark after passivation, or by performing a 10 min pre-anneal at 200 °C [39] prior to lifetime measurement. The samples were subjected to ~50 close-up flashes from the lifetime tester to dissociate FeB pairs [40], after which an initial lifetime measurement was made immediately. It is noted that the aggregated illumination time of the flashes of light used to dissociate the iron–boron pairs is very short (<20 ms), so any effect on the formation of boron–oxygen defects is kept to a minimum [39]. A second lifetime measurement was made more than 24 h later, which was sufficient time to reassociate the FeB pairs [19], [34] and [41]. The two lifetime measurements are then analysed to give the concentration of iron that exists in FeB pairs using an established method [30],[40] and [42]. This concentration is henceforth referred to as the *bulk* iron concentration, and excludes iron present in other forms such as iron silicide precipitates, or iron bound to, or precipitated at, oxide precipitates and any surrounding defects. The specific analysis approach used is described in a previous publication [19]. The essential feature is that SRH statistics (Eq. (1)) are used with the recombination parameters of Rein and Glunz [30] to determine the bulk iron concentration required to account for a lifetime change at a given injection level. For the results presented in this paper, the injection level used was $0.2p_0$. The bulk iron concentrations in the "uncontaminated" samples were always $\leq 1.5\times10^{12}$ cm$^{-3}$.

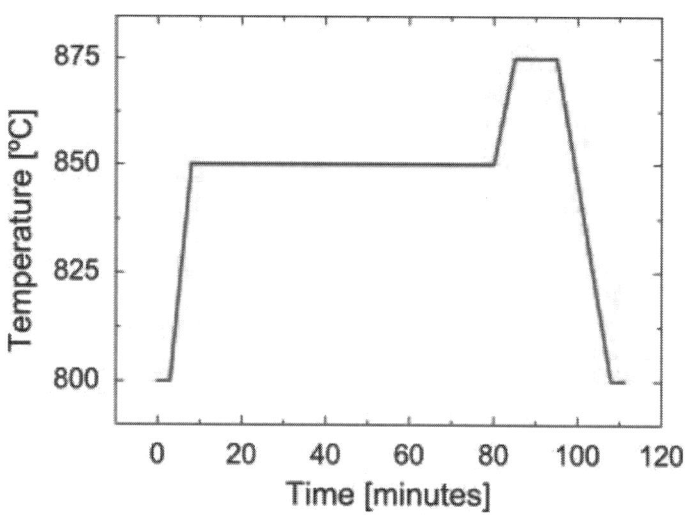


**Figure 2.** The time-temperature profile used in the phosphorus diffusion gettering process for which the lifetime data are shown in Fig. 7.

We express our data in terms of a "residual" minority carrier lifetime, $\tau_{residual}$, defined according to:

$$\frac{1}{\tau_{\text{residual}}} = \frac{1}{\tau_{\text{measured}}} - \left( \frac{1}{\tau_{\text{band-to-band}}} + \frac{1}{\tau_{\text{CEAuger}}} + \frac{1}{\tau_{\text{Fe}_i}} \right) \quad (7)$$

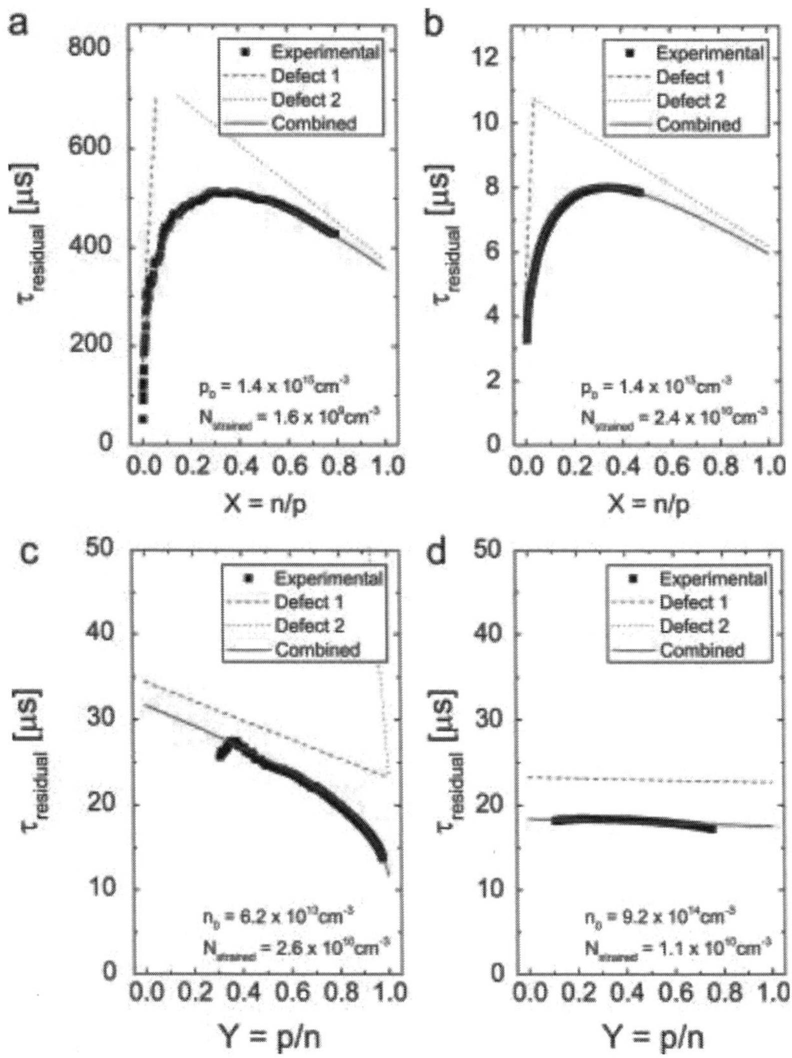


**Figure 3.** Residual minority carrier measured in uncontaminated samples containing oxide precipitates. Graphs (a) and (b) are for p-type material. Dislocations and stacking faults were found to surround the precipitates in (b), but not in (a). Graphs (c) and (d) are for n-type material with different doping levels.

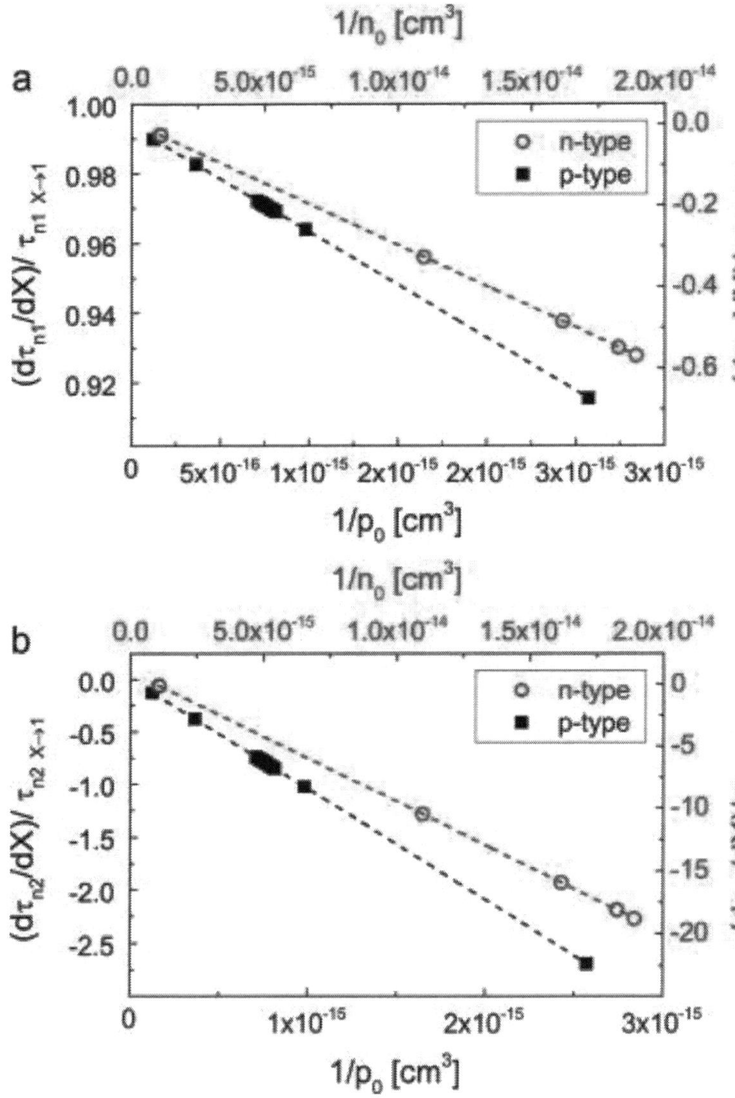


**Figure 4.** Plots used to deduce the values of: (a) $Q_1$, $Q_1n_1+p_1$ (for p-type) and $n_1+p_1/Q_1$ (for n-type) for Defect 1; and (b) $Q_2$, $Q_2n_2+p_2$ (for p-type) and $n_2+p_2/Q_2$ (for n-type). The p-type plots are in accordance with Eq. (5). Each data point corresponds to one sample.

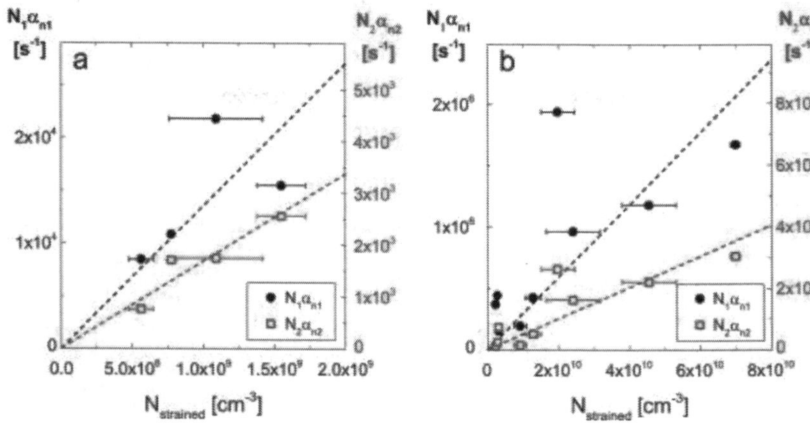

**Figure 5.** Dependence of the *Nαn* fit parameters on density of strained oxide precipitates for Defect 1 and Defect 2 in p-type samples. Graph (a) is for samples in which oxide precipitates are not surrounded by dislocations and stacking faults. Graph (b) is for samples in which oxide precipitates are surrounded by dislocations, and, in some cases, stacking faults. Density error bars represent the standard deviation of counting multiple regions of etch pits.

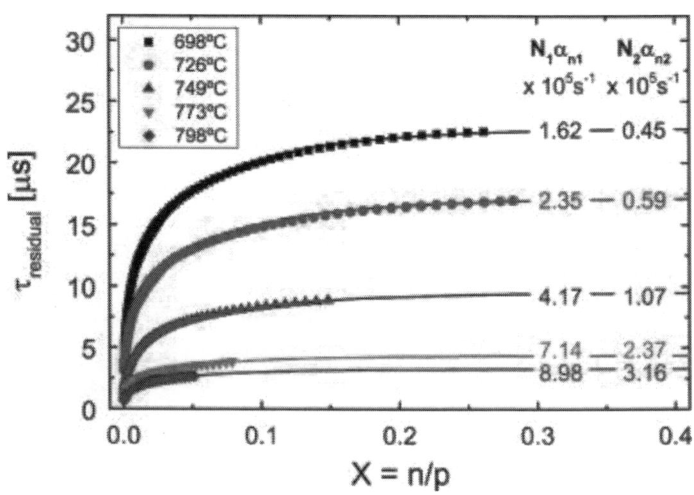

**Figure 6.** Residual minority carrier lifetime in p-type silicon containing oxide precipitates contaminated with iron at the temperatures shown. These data are for samples taken from the same wafer, which had a doping level of ~$8\times10^{15}$ $cm^{-3}$ and a strained oxide precipitate density of 1 $cm^{-3}$ to $2\times10^{9}$ $cm^{-3}$.

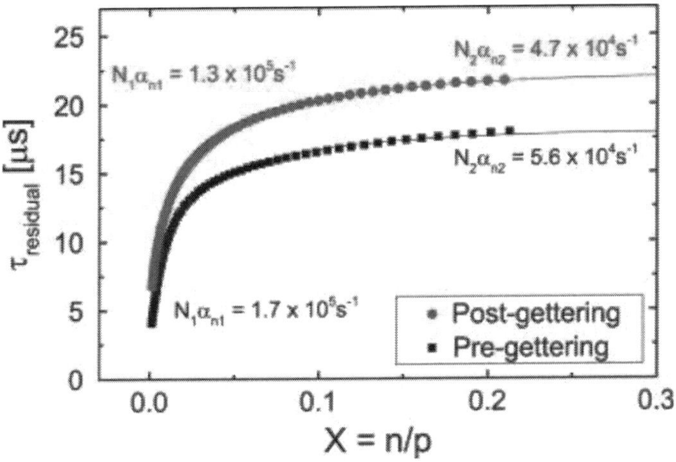


**Figure 7.** Residual minority carrier lifetime before and after phosphorus diffusion gettering in a p-type silicon sample with an oxide precipitate concentration of $1.2\times10^{9}$ $cm^{-3}$ and doping level of $8\times10^{15}$ $cm^{-3}$. The bulk iron concentration was $1.4\times10^{12}$ $cm^{-3}$ before gettering, and $5.2\times10^{11}$ $cm^{-3}$ after gettering.

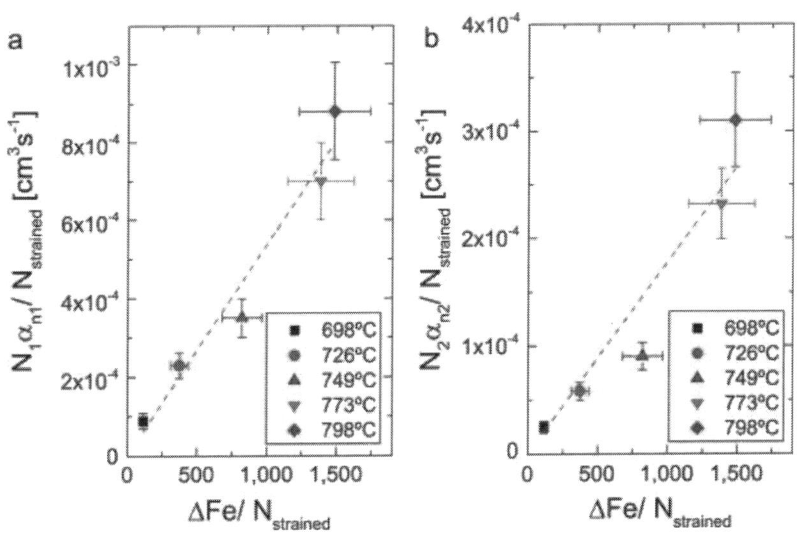


**Figure 8.** The SRH parameter *Nan* per strained precipitate for both defects plotted against the iron lost from the bulk given by Eq. (8) per precipitate. The data are from the same samples for which injection-dependent lifetime measurements are shown in Fig. 6.

where $\tau_{measured}$ is the measured minority carrier lifetime with iron in the interstitial state, $\tau_{band\text{-}to\text{-}band}$ is the lifetime due to band-to-band recombination (from [43]), $\tau_{CE\ Auger}$ is the lifetime due to Coloumb-enhanced Auger recombination (from [44]) and $\tau_{Fei}$ is the lifetime due to SRH recombination at bulk interstitial iron (parameters from [30]). More details of these corrections are given in our previous paper [19]. It is not possible to measure the interstitial iron concentration in n-type samples, so no correction is made for bulk iron-related recombination in such samples, and, besides which, the recombination activity of iron in n-type silicon is generally much less significant than in p-type silicon [45]. The effects of any remaining surface recombination are not factored out of the injection-dependent lifetime data. This is not believed to have a substantial effect on our findings. The work of Aberle et al. shows the surface recombination velocity to be dependent on injection-level, with the surface recombination velocity increasing at lower levels of injection [29]. However, this alone is insufficient to explain the injection-level dependence we observe. It is noted that the same fit parameters can be used to fit lifetime data from samples with a high bulk lifetime (low precipitate density) and low bulk lifetime (iron contaminated precipitates with a high density), which suggests that differences between surface recombination rates between samples are not substantial.

## 3.3. Phosphorus diffusion gettering

Some uncontaminated samples were subjected to a phosphorus diffusion gettering process. The injection-dependent minority carrier lifetime and the bulk iron concentration were first measured as above. The silicon nitride was then removed from both surfaces by HF (40%) for up to ~15 s at room temperature. Samples were then RCA cleaned, followed by a phosphorus diffusion gettering process in a standard quartz-tube furnace with the temperature profile shown in Fig. 2. The process used is the standard $POCl_3$-based process for formation of high efficiency passivated emitter and rear cell (PERC) type solar cells at ISFH. The phosphorus glass was then removed by immersion in HF (40%) for 3 min at room temperature, and the emitter was removed by a KOH etch for 3 min at ~90 °C. This removed a total of ~10 μm of material from the surfaces, reducing the wafer thickness from ~700 μm to ~690 μm. The amount of material removed was calculated by weighing the wafer before and afterwards. Samples were then subjected to another RCA clean followed by silicon nitride surface passivation and lifetime measurement, as described above.

# 4. RESULTS

## 4.1. "Uncontaminated" samples

In Fig. 3, the residual minority carrier lifetime measured in uncontaminated samples containing oxide precipitates is plotted against $X=n/p$ for two p-type specimens and$Y=p/n$ for two n-type specimens. The supporting TEM study [3] found dislocations and stacking faults to surround the oxide precipitates in one of the p-type sample types, but not the other. The two n-type samples have different doping levels.

Similar curves to those in Fig. 3(a) and (b) were obtained in ~100 other p-type samples studied. Eq. (1) shows that a single SRH centre gives a lifetime response which is linear with $X=n/p$, but the variation obtained experimentally is clearly not linear. Similarly Fig. 3(c) shows clear non-linearity with $Y=p/n$ in the n-type case too. Thus it is the case that oxygen precipitation introduces more than simple single-level SRH recombination centre. We have previously ruled out the possibility that precipitation introduces a single defect with two energy levels [21]. We find in all cases that the injection-dependence of the lifetime can be fitted with two single-levels that act independently, i.e. $(1/\tau_{residual})=((1/\tau_1)+(1/\tau_2))$. These are referred to simply as "Defect 1" and "Defect 2", and their individual and combined effects that best fit the experimental data are plotted on the graphs in Fig. 3.

To determine the properties of the two defects, the same approach as in the example shown in Fig. 1 was used. The gradients of the fits to the experimental data for both defects for a large number of p-type samples were normalised by the high injection ($X \rightarrow 1$) limits and were plotted against $1/p_0$ according to Eq. (5). A similar approach was taken for the n-type samples (see Ref. [21]). The plots are shown in Fig. 4. These plots were used to deduce the value of $Q$ from the intercept and the values of $Qn_1+p_1$ (for p-type) and $n_1+p_1/Q$ (for n-type) from the gradient. For Defect 1, both the p-type and n-type data give the ratio of the capture coefficient for electrons ($\alpha n_1$) to that of holes ($\alpha p_1$) as$Q_1=\alpha n_1/\alpha p_1=157$. For Defect 2, both the p-type and the n-type data give the ratio of the capture coefficient for holes ($\alpha p_2$) to that of electrons ($\alpha n_2$) for Defect 2, $1/Q_2=\alpha p_2/\alpha n_2=1{,}200$. For Defect 1, for which $n_1$ and $p_1$ are the SRH densities given by Eqs.(2) and (3), respectively, the parameter $Q_1n_1+p_1$ is $4.8\times10^{15}$ cm$^{-3}$ from the p-type data and the parameter $n_1+(p_1/Q_1)$ is $3.1\times10^{13}$ cm$^{-3}$ from the n-type data. For Defect 2, for which $n_2$ and $p_2$ are the corresponding SRH densities, the parameter $Q_2n_2+p_2$ is $1.0\times10^{15}$ cm$^{-3}$ from the p-type and the parameter

$n_2+(p_2/Q_2)$ is $1.3\times10^{18}$ cm$^{-3}$ from the n-type data. Temperature-dependent lifetime data reported previously [21] were used to determine the half of the band-gap in which the energy levels of the two defects lie. We deduce that Defect 1 is at $EV$ +0.22 eV, and that Defect 2 is at $EC$−0.08 eV. The SRH parameters are summarised in Table 1. The temperature dependent lifetime data show that the capture coefficient for holes at Defect 1 ($\alpha p_1$) decreases with temperature with a 0.20 eV activation energy, and that the capture coefficient for electrons at Defect 2 ($\alpha n_2$) decreases with temperature with a 0.14 eV activation energy [21].

**Table 1.** SRH parameters for recombination at oxide precipitates in silicon deduced from both p-type and n-type data. Also included are the $N\alpha n$ per strained precipitate values deduced from a best fit to data for uncontaminated p-type samples presented in Fig. 5.

| | $E_T$ (eV) | $Q=\alpha_n/\alpha_p$ | $N\alpha_n/N_{strained}$ (no surrounding dislocations or stacking faults) (cm$^3$ s$^{-1}$) | $N\alpha_n/N_{strained}$ (with surrounding dislocations and stacking faults) (cm$^3$ s$^{-1}$) |
|---|---|---|---|---|
| Defect 1 | $E_V$+0.22 | 157 | $1.7\times10^{-5}$ | $2.9\times10^{-5}$ |
| Defect 2 | $E_C$−0.08 | $8.33\times10^{-4}$ | $1.8\times10^{-6}$ | $5.1\times10^{-6}$ |

As well as extracting SRH parameters from the lifetime data, it is possible to extract information on the density of each defect and to correlate this with the precipitate density measured by chemical etching. It is not possible to determine the absolute densities of Defect 1 ($N_1$) and Defect 2 ($N_2$) from lifetime measurements alone. However, using Eq.(6) it is possible to deduce $N_1\alpha n_1$ and $N_2\alpha n_2$ for each lifetime measurement in p-type material. A similar approach for n-type material allows for $N_1\alpha p_1$ and $N_2\alpha p_2$ to be extracted. The p-type parameters are plotted against the measured concentration of strained precipitates in Fig. 5(a) for the case when oxide precipitates are not surrounded by dislocations and stacking faults and in Fig. 5(b) in the case when the oxide precipitates are surrounded by other extended defects. In both plots, the correlation of $N_1\alpha n_1$ and$N_2\alpha n_2$ with $N_{strained}$ is approximately linear. The gradients of the linear fit shown are given alongside the SRH parameters in Table 1. Similar plots for n-type material have been published previously [21]. We note that the relationship is only approximately linear, and we suggest this is due to differences in the decoration of the precipitates by impurities, as discussed in the next section.

## 4.2. Iron contaminated samples

Injection-dependent lifetime measurements on a set of p-type iron contaminated samples are plotted in Fig. 6. The samples used all came from the same wafer (so have very similar precipitate densities), and different levels of iron were diffused into the samples by varying the contamination temperature. The residual lifetime reduces with contamination temperature. It is noted that the lifetime component associated with bulk iron is factored out of $\tau_{residual}$ (Eq. (7)), so reductions in lifetime are assumed to be due to recombination associated with iron segregated to oxide precipitates and any surrounding defects (dislocations and stacking faults).

It is remarkable that the injection response of the residual lifetime has the same form in iron-contaminated and "uncontaminated" samples. In fact, the same defect parameters can be used to fit the experimental data in both cases. $N\alpha n$ terms for both Defect 1 and Defect 2 have been extracted from fits to the data in Fig. 6. The values of these fit parameters increase with contamination temperature, and the ratio of $N_1\alpha n_1$ to $N_2\alpha n_2$ is fairly consistent at between 2.8 and 4.0. We discuss the relationship between recombination activity and iron decoration of the oxide precipitates and surrounding defects in Section 5.2.

## 4.3. Samples subjected to phosphorus diffusion gettering

Fig. 7 shows residual minority carrier lifetime plotted against $X=n/p$ for a typical sample before and after phosphorus diffusion gettering using the temperature profile shown inFig. 2. The injection response before and after gettering can be fitted using the parameters for Defect 1 and Defect 2 shown in Table 1. The gettering process increases the overall lifetime, and the $N\alpha n$ term for both defects is reduced by ~20%. The gettering process reduces the bulk iron concentration by a factor of ~2.7.

# 5. DISCUSSION

## 5.1. Parameterisation in terms of SRH statistics

By studying the injection-dependence of minority carrier lifetime as a function of doping level in both p-type and n-type silicon it has been possible to quantify recombination at oxide precipitates in terms of SRH statistics. The recombination processes can be parameterised in terms of two independent single-level SRH centres, the parameters for which are stated in Table 1. It is not possible to determine the absolute density of these centres from our lifetime measurements alone. However, we have provided an empirical correlation between the strained oxide precipitate density (determined by etching) and$N\alpha_n$ for each defect. The existence of dislocations/stacking faults around the precipitates was found to increase the $N\alpha_n$ per precipitate values by a factor of ~2 to 3. These other defects did not however introduce any other detectable recombination centres.

It is interesting to note that the same parameterisation of the recombination activity can be used before and after phosphorus diffusion gettering (Fig. 7). This rules out the possibility that iron silicide precipitates form in the bulk during gettering. It is likely that these would have different recombination parameters, and our results show that the gettering process does not introduce other recombination centres in detectable concentrations for the conditions studied.

## 5.2. The role of iron in the recombination process

The data presented in Fig. 6 shows that the residual lifetime in iron-contaminated samples can be parameterised using the same two defects found in the "uncontaminated" samples. Iron contamination does not change the energy levels and the ratio of capture coefficients required to fit the injection-dependent lifetime data. However, the $N\alpha_n$ parameters required to fit the lifetime data increased with increasing contamination temperature. Although iron solubility is clearly a factor in the resultant recombination activity, it is not the only one, as we have previously shown that the cooling rate after the contamination anneal also plays a role [23]. The bulk iron concentrations in the contaminated samples measured by photodissociation of FeB pairs are usually considerably lower than the solubility values [23]. We assume that the difference in concentrations ($\Delta Fe$) has segregated to the oxide precipitates according to:

$$\Delta Fe=[Fe_{solubility}]-[Fe_{bulk}] \tag{8}$$

where $[Fe_{solubility}]$ is the solubility of iron at the contamination temperature used, and $[Fe_{bulk}]$ is the bulk iron concentration in the contaminated sample measured by photodissociation of FeB pairs. The iron solubility in the relevant temperature range is given by

$$[Fe_{solubility}] = 1.3 \times 10^{21}\exp\left(-\frac{1.8\text{eV}}{kT}\right)\text{cm}^{-3}$$

where $T$ is the contamination temperature [34] and [35].

Fig. 8 shows the relationship between the $\alpha nN$ parameter per strained precipitate and $\Delta Fe$ per strained precipitate for the five samples for which the lifetime data are plotted inFig. 6. This relationship is approximately linear for Defect 1 and Defect 2. This, combined with previously published data on samples with a wider variety of precipitate concentrations [23], suggests that segregated iron determines the recombination activity of the strained oxide precipitates and surrounding defects. On the assumption that $\alpha n$ is invariant, the density of the recombination centres appears to be proportional to the number of iron atoms segregated to the precipitates. This implies that atomic decoration is more likely than iron precipitation at the oxide precipitates.

It is known that oxygen precipitation creates so-called *Pb* dangling bonds [20] and [46]which are known to give rise to recombination activity [20]. However, these can be passivated by hydrogen [47] and [48], which is likely to be introduced into our samples during the PECVD surface passivation step [49]. It is therefore possible that the only recombination activity we detect (and that would be expected in a completed solar cell) is due to impurities segregated to oxide precipitates and surrounding defects. We can estimate the number of iron atoms required to explain the recombination activity we measured in so-called "uncontaminated" samples. Fig. 5 shows the recombination rate via both defects in "uncontaminated" samples varies approximately linearly with precipitate density, with $N_1\alpha n_1/N_{strained}=(1.7\rightarrow 2.9)\times 10^{-5}$ cm$^3$ s$^{-1}$ and $N_2\alpha n_2/N_{strained}=(1.8\rightarrow 5.1)\times 10^{-6}$ cm$^3$ s$^{-1}$ (Table 1). Assuming the state density to be proportional to $\Delta Fe$, the linear relationship shown in Fig. 8 is consistent with each strained precipitate in "uncontaminated" samples being decorated with ≤50 atoms of iron. The idea that very low levels of impurity contamination can have substantial effects on the electrical properties of extended defects is not a new one. It has been previously shown that the recombination activity of dislocation is strongly enhanced by contamination with metallic impurities [50] and [51].

It would not be surprising that oxide precipitates in monocrystalline Cz-Si solar wafers are decorated by such low levels of metal impurities. The iron concentration at the precipitates required for the electrical activity seen in "uncontaminated" samples is much less than the iron solubility at temperatures at used in solar cell processing. Haunschild et al. found a 10 s thermal treatment at 800 °C to result in a strong enhancement of the recombination activity of rings of oxide precipitates. Their thermal process would have been sufficient to redistribute low concentrations of bulk iron in the material, resulting in increased iron decoration of the precipitates. The efficiency reductions of up to 4% (absolute) that can arise due to oxygen precipitation in Cz-Si [7], is therefore likely to involve transition metal contamination of the precipitates and surrounding defects.

It is interesting to note that the recombination activity of oxide precipitates in both "uncontaminated" and contaminated samples appears to be approximately dependent upon precipitate *density* and not size. An explanation for this is that iron segregates to regions of the precipitates whose number is invariant with size, such as in the vicinity of precipitate corners.

## 5.3. Extended versus point-like defects

It is interesting that although oxide precipitates and surrounding dislocations and stacking faults are *extended* defects, we find that their effect on bulk minority carrier lifetime is that expected from *point-like* defects. Our measurements were obtained in a wide range of samples (both types, different doping levels, a wide range of precipitate densities of morphologies, with/without iron contamination) and previously at different temperatures [21]. Using our lifetime study alone, we have yet to find any feature of the process which cannot be explained using SRH statistics. We note that others have applied a SRH-like approach to this problem in the past [52] and [53]. However, we also acknowledge that other previous studies have invoked forms of barrier-controlled capture at the precipitates [54]. Reconciliation of the two approaches is a topic that requires further study.

## 5.4. Lifetime in n-type and p-type silicon

There has been considerable recent interest in producing photovoltaics from n-type silicon substrates [55]. This is at partly because n-type material can have very high carrier lifetimes [56] due to the absence of light-induced degradation in boron-free material [57], and reduced recombination at interstitial iron [30] and [45]. Do oxide precipitates and associated defects affect recombination in one type more than another? The parameterisation established here and previously [21] and [23] can be used to estimate this.

We have established that iron decoration of oxide precipitates strongly enhances the recombination activity, so for a fair comparison we need to account for the iron decoration. Fitting the injection-dependent lifetime data alone does not allow for the separation of the state density ($N$) from the capture coefficient for electrons or holes ($\alpha n$or $\alpha p$, respectively). However, from assuming all the recombination activity arises due to iron decoration, we have previously combined iron-loss measurements with the lifetime data to estimate $\alpha n_1$=2×10$^{-6}$ cm$^3$ s$^{-1}$ and $\alpha n_2$=4×10$^{-7}$ cm$^3$ s$^{-1}$[23]. This assumes that each iron atom at an oxide precipitate gives rise to one of each of Defect 1 and Defect 2. We can use the values of $Q$ in Table 1 to estimate $\alpha p_1$=1.3×10$^{-8}$ cm$^3$ s$^{-1}$ and$\alpha p_2$=4.8×10$^{-4}$ cm$^3$ s$^{-1}$. Under the stated assumptions, this gives us the full parameterisation necessary to estimate carrier lifetime due to iron-decorated oxide precipitates in p-type and n-type silicon.

A plot of minority carrier lifetime multiplied by state density ($N$=$N_1$=$N_2$) versus excess minority carrier concentration in n-type and p-type silicon with different doping levels is shown in Fig. 9. Under the assumptions stated above, the state density is the density of iron atoms at the oxide precipitates. The estimated residual minority carrier lifetime depends strongly upon doping level and injection level, with the lifetimes at high injection tending to the ambipolar lifetime. At typical substrate doping levels for photovoltaics (~10$^{16}$ cm$^{-3}$), the lifetime in p-type material is higher than in n-type for injection levels at which the lifetime is not at the ambipolar limit. For higher substrate doping levels the lifetime benefit of p-type over n-type becomes more substantial. For lower substrate doping levels, the lifetime values are similar in the p-type and n-type cases, with any predicted differences likely to lie within the errors of the SRH parameters. In summary, the recombination activity associated with just the (iron-decorated) oxide precipitates is slightly better in typical p-type PV substrates than in n-type. That said, the increased recombination activity at bulk interstitial iron in p-type material [30] and [45] could easily eliminate any possible advantage.

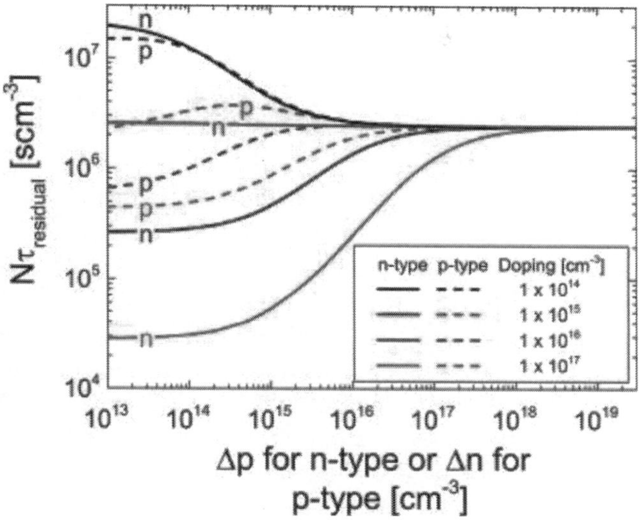


**Figure 9.** Simulated product of state density and residual minority carrier lifetime versus excess minority carrier density in p-type and n-type silicon. The simulation uses the values of *ET* and *Q* in Table 1, and assumes$\alpha n_1=2\times10^{-6}$ cm$^3$ s$^{-1}$ and $\alpha n_2=4\times10^{-7}$ cm$^3$ s$^{-1}$. Temperature is taken as 25 °C.

## 5.5. Competition between gettering systems

In silicon wafers for integrated circuits oxide precipitates and associated defects are frequently used as a way of internally gettering transition metal impurities, thus preventing the impurities harming the devices [5] and [6]. We performed phosphorus diffusion gettering on samples containing oxide precipitates, which enables us to assess the relative strength of the two gettering systems. For the results shown in Fig. 7, the bulk iron concentration was $1.4\times10^{12}$ cm$^{-3}$ before gettering and $5.2\times10^{11}$ cm$^{-3}$ after gettering, and *Nαn* terms associated with the oxide precipitates fell by ~20%. As we have shown that the *Nαn* parameters correlate with the iron lost to the precipitates (Fig. 8), we conclude that the bulk iron lost has not been gettered to the oxide precipitates, and has most likely been gettered to the phosphorus diffused layer. Thus, for the conditions investigated, phosphorus diffusion gettering is more effective at gettering bulk iron than oxide precipitates. This is very encouraging from the perspective of solar wafers containing oxide precipitates and suggests that the recombination activities will not increase (and may decrease slightly) during the phosphorus diffusion gettering process. It is also noted that our findings are consistent with those of Rinio et al., who found evidence to suggest that impurities are gettered preferentially to the emitter rather than native crystal defects in mc-Si [58].

The correlation between iron loss and recombination activity shown in Fig. 8 implies that our analysis approach is capable of measuring very low relative concentrations of atomic iron at oxide precipitates. The results presented in Fig. 7 show that the states associated with oxide precipitates are reduced by ~20% by the phosphorus diffusion gettering process used. An explanation for this reduction is that this proportion of iron has dissociated from the oxide precipitates and surrounding defects. The future focus of our research is to use this approach with different thermal conditions to understand the fundamental thermodynamics of the interaction between iron and oxide precipitates.

## 6. CONCLUSIONS

As a consequence of using fast crystal pulling rates, some monocrystalline Cz-Si solar wafers contain vacancy-rich regions which act as preferential sites for precipitation of supersaturated bulk oxygen. Oxide precipitates and surrounding defects are associated with recombination activity, and this can have a substantial detrimental impact on the conversion efficiency of solar cells. We have used injection-dependent minority carrier lifetime measurements to study this recombination activity in ~100 samples with different types, doping levels, precipitate morphologies and sizes. In all cases we find the lifetime can be parameterised in terms of two independent defect states. The first defect is at$E_V$+0.22 eV and has a capture coefficient for electrons ~157 times greater than that for holes. The second defect is at $E_C$−0.08 eV and has a capture coefficient for holes ~1200 times greater than that for electrons. The concentration of centres appears to scale with the density (not size) of the precipitates, with surrounding dislocations and stacking faults increasing the density of centres per precipitate.

Remarkably, we find the same defect states in oxide precipitate-containing samples which have been intentionally contaminated with iron. The density of the states per precipitate is dependent on the iron loss from the bulk per precipitate. We therefore conclude that at least some of the recombination activity of oxide precipitates in "uncontaminated" samples is due to the segregation of iron to the precipitates and surrounding defects. We find that a phosphorus diffusion gettering process reduces the density of states associated with the oxide precipitates by ~20%, and that bulk iron is preferentially gettered to the phosphorus diffused layer rather than to the oxide precipitates.

## ACKNOWLEDGEMENTS

The authors thank D. Gambaro, M. Cornara, and M. Olmo (SunEdison) for performing precipitation treatments and characterization, R. Chakalova (University of Oxford) for assistance with sample cleaning and surface passivation, and P.R. Wilshaw (University of Oxford) for helpful discussions. JDM is the holder of a Royal Academy of Engineering/EPSRC Research Fellowship and an EPSRC First Grant (EP/J01768X/1).

## REFERENCES

1. Borghesi, B. Pivac, A. Sassella, A. Stella, Oxygen precipitation in silicon, Journal of Applied Physics 77 (1995) 4169.

2. W. Bergholz, M.J. Binns, G.R. Booker, J.C. Hutchison, S.H. Kinder, S. Messoloras, R.C. Newman, R.J. Stewart, J.G. Wilkes, A study of oxygen precipitation in silicon using high-resolution transmission electron microscopy, small-angle neutron scattering and infrared absorption, Philosophical Magazine B 59 (1989) 499.

3. R. Falster, V.V. Voronkov, V.Y. Resnik, M.G. Milvidskii, Thresholds for effective internal gettering in silicon wafers, Proceedings of the Electrochemical Society, High Purity Silicon, VIII, 2004, pp. 188–201.

4. R.C. Newman, Oxygen diffusion and precipitation in Czochralski silicon, Journal of Physics: Condensed Matter 12 (2000) R335.

5. S.M. Myers, M. Seibt, W. Schröter, Mechanisms of transition-metal gettering in silicon, Journal of Applied Physics 88 (2000) 3795.

6. R.J. Falster, W. Bergholz, The gettering of transition metals by oxygen-related defects in silicon, Journal of the Electrochemical Society 137 (1990) 1548.

7. J. Haunschild, I.E. Reis, J. Geilker, S. Rein, Detecting efficiency-limiting defects in Czochralski-grown silicon wafers in solar cell production using photoluminescence imaging, Physica Status Solidi Rapid Research Letters 5 (2011) 199.

8. P.K. Kulshreshtha, Y. Yoon, K.M. Youssef, E.A. Good, G. Rozgonyi, Oxygen precipitation related stress-modified crack propagation in high growth rate Czochralski silicon wafers, Journal of the Electrochemical Society 159 (2012) H125.

9. R. Søndena, Y. Hu, M. Juel, M.S. Wiig, H. Angelskår, Characterization of the OSFband structure in n-type Cz-Si using photoluminescence-imaging and visual inspection, Journal of Crystal Growth 367 (2013) 68.

10. K. Youssef, M. Shi, C. Radue, E. Good, G. Rozgonyi, Effect of oxygen and associated residual stresses on the mechanical properties of high growth rate Czochralski silicon, Journal of Applied Physics 113 (2013) 133502.

11. H.J. Möller, C. Funke, A. Lawerenz, S. Riedel, M. Werner, Oxygen and lattice distortions in multicrystalline silicon, Solar Energy Materials & Solar Cells 72 (2002) 403. [12] K. Bothe, K. Ramspeck, D. Hinken, C. Schinke, J. Schmidt, S. Herlufsen, R. Brendel, J. Bauer, J.-M. Wagner, N. Zakharov, O. Breitenstein, Luminescence emission from forward- and reverse-biased multicrystalline silicon solar cells, Journal of Applied Physics 106 (2009) 104510.

12. M. Tajima, Y. Iwata, F. Okayama, H. Toyota, H. Onodera, T. Sekiguchi, Deep-level photoluminescence due to dislocations and oxygen precipitates in multicrystalline Si, Journal of Applied Physics 111 (2012) 113523.

13. V.V. Voronkov, R. Falster, Grown-in microdefects, residual vacancies and oxygen precipitation bands in Czochralski silicon, Journal of Crystal Growth 204 (1999) 462.

14. V.V. Voronkov, The mechanism of swirl defects formation in silicon, Journal of Crystal Growth 59 (1982) 625.

15. L. Chen, X. Yu, P. Chen, P. Wang, X. Gu, J. Lu, D. Yang, Effect of oxygen precipitation on the performance of Czochralski silicon solar cells, Solar Energy Materials & Solar Cells 95 (2011) 3148.

16. W. Seifert, M. Kittler, M. Seibt, A. Buczkowski, Contrastive recombination behaviour of metal silicide and oxygen precipitates in n-type silicon: attempt at an explanation, Solid State Phenomena 47–48 (1996) 365–370.

17. F.G. Kirscht, Y. Furukawa, W. Seifert, K. Schmalz, A. Buczkowski, S.B. Kim, H. Abe, H. Koya, J. Bailey, Electrical characteristics of oxygen precipitation related defects in Czochralski silicon wafers, Materials Science and Engineering B 36 (1996) 230–236.

18. J.D. Murphy, K. Bothe, M. Olmo, V.V. Voronkov, R.J. Falster, The effect of oxide precipitates on minority carrier lifetime in p-type silicon, Journal of Applied Physics 110 (2011) 053713.

19. V. Lang, J.D. Murphy, R.J. Falster, J.J.L. Morton, Spin-dependent recombination in Czochralski silicon containing oxide precipitates, Journal of Applied Physics 111 (2012) 013710.

20. J.D. Murphy, K. Bothe, R. Krain, V.V. Voronkov, R.J. Falster, Parameterisation of injection-dependent lifetime measurements in semiconductors in terms of Shockley–Read–Hall statistics: an application to oxide precipitates in silicon, Journal of Applied Physics 111 (2012) 113709.

21. K. Bothe, R.J. Falster, J.D. Murphy, Room temperature sub-bandgap photoluminescence from silicon containing oxide precipitates, Applied Physics Letters 101 (2012) 032107.

22. J.D. Murphy, K. Bothe, V.V. Voronkov, R.J. Falster, On the mechanism of recombination at oxide precipitates in silicon, Applied Physics Letters 102 (2013) 042105.

23. W. Shockley, W.T. Read, Statistics of the recombinations of holes and electrons, Physical Review 87 (1952) 835.

24. R.N. Hall, Electron–hole recombination in germanium, Physical Review 87 (1952) 387.

25. D. Macdonald, A. Cuevas, Validity of simplified Shockley–Read–Hall statistics for modeling carrier lifetimes in crystalline silicon, Physical Review B: Condensed Matter 67 (2003) 075203.

26. S. Rein, T. Rehrl, W. Warta, S.W. Glunz, Lifetime spectroscopy for defect characterization: systematic analysis of the possibilities and restrictions, Journal of Applied Physics 91 (2002) 2059.

27. D. Macdonald, A. Cuevas, Trapping of minority carriers in multicrystalline silicon, Applied Physics Letters 74 (1999) 1710.

28. A.G. Aberle, T. Lauinger, J. Schmidt, R. Hezel, Injection-level dependent surface recombination velocities at the silicon-plasma silicon nitride interface, Applied Physics Letters 66 (1995) 2828.

29. S. Rein, S.W. Glunz, Electronic properties of interstitial iron and iron–boron pairs determined by means of advanced lifetime spectroscopy, Journal of Applied Physics 98 (2005) 113711.

30. J. Schmidt, Effect of dissociation of iron–boron pairs in crystalline silicon on solar cell properties, Progress in Photovoltaics: Research and Applications 13 (2005) 325.

31. H. Nagel, B. Lenkeit, W. Schmidt, Fill factor losses due to injection-level dependent bulk lifetimes in crystalline silicon solar cells, in: 20th European Photovoltaic Solar Energy Conference, Barcelona, Spain, 2005, pp. 1271–1274.

32. M.A. Green, Intrinsic concentration, effective densities of states, and effective mass in silicon, Journal of Applied Physics 67 (1990) 2944.

33. J.D. Murphy, R.J. Falster, Contamination of silicon by iron at temperatures below 800 1C, Physica Status Solidi Rapid Research Letters 5 (2011) 370.

34. J.D. Murphy, R.J. Falster, The relaxation behaviour of supersaturated iron in single-crystal silicon at 500–750 1C, Journal of Applied Physics 112 (2012) 113506.

35. T. Lauinger, J. Moschner, A.G. Aberle, R. Hezel, Optimization and characterization of remote plasma-enhanced chemical vapor deposition silicon nitride for the passivation of p-type crystalline silicon surfaces, Journal of Vacuum Science & Technology A 16 (1998) 530.

36. P. Karzel, P. Frey, S. Fritz, G. Hahn, Influence of hydrogen on interstitial iron concentration in multicrystalline silicon during annealing steps, Journal of Applied Physics 113 (2013) 114903.

37. R.A. Sinton, A. Cuevas, Contactless determination of current–voltage characteristics and minority-carrier lifetimes in semiconductors from quasi-steadystate photoconductance data, Applied Physics Letters 69 (1996) 2510.

38. K. Bothe, J. Schmidt, Electronically activated boron–oxygen-related recombination centers in crystalline silicon, Journal of Applied Physics 99 (2006) 013701.

39. G. Zoth, W. Bergholz, A fast, preparation-free method to detect iron in silicon, Journal of Applied Physics 67 (1990) 6764.

40. W. Wijaranakula, The reaction kinetics of iron–boron pair formation and dissociation in p-type silicon, Journal of the Electrochemical Society 140 (1993) 275.

41. D.H. Macdonald, L.J. Geerligs, A. Azzizi, Iron detection in crystalline silicon by carrier lifetime measurements for arbitrary injection and doping, Journal of Applied Physics 95 (2004) 1021.

42. H. Schlangenotto, H. Maeder, W. Gerlach, Temperature dependence of the radiative recombination coefficient in silicon, Physica Status Solidi A Applications and Material Science 21 (1974) 357.

43. M.J. Kerr, A. Cuevas, General parameterization of Auger recombination in crystalline silicon, Journal of Applied Physics 91 (2002) 2473.

44. D. Macdonald, L.J. Geerligs, Recombination activity of interstitial iron and other transition metal point defects in p- and n-type crystalline silicon, Applied Physics Letters 85 (2004) 4061.

45. M. Koizuka, H. Yamada-Kaneta, Electron spin resonance centers associated with oxygen precipitates in Czochralski silicon crystals, Journal of Applied Physics 88 (2000) 1784.

46. M. Koizuka, H. Yamada-Kaneta, Gap states caused by oxygen precipitation in Czochralski silicon crystals, Journal of Applied Physics 84 (1998) 4255.

47. E. Cartier, J.H. Stathis, D.A. Buchanan, Passivation and depassivation of silicon dangling bonds at the Si/SiO2 interface by atomic hydrogen, Applied Physics Letters 63 (1993) 1510.

48. A.G. Aberle, Overview on SiN surface passivation of crystalline silicon solar cells, Solar Energy Materials and Solar Cells 65 (2001) 239–248.

49. V. Kveder, M. Kittler, W. Schröter, Recombination activity of contaminated dislocations in silicon: a model describing electron-beam-induced current contrast behavior, Physical Review B: Condensed Matter 63 (2001) 115208.

50. T.S. Fell, P.R. Wilshaw, M.D. de Coteau, EBIC investigations of dislocations and their interactions with impurities in silicon, Physica Status Solidi A Applications and Material Science 138 (1993) 695.

51. J. Vanhellemont, E. Simoen, A. Kaniava, M. Libezny, C. Claeys, Impact of oxygen related extended defects on silicon diode characteristics, Journal of Applied Physics 77 (1995) 5669.

52. W. Seifert, M. Kittler, J. Vanhellemont, EBIC study of recombination activity of oxygen precipitation related defects in Si, Materials Science and Engineering B 42 (1996) 260.

53. J.M. Hwang, D.K. Schroder, Recombination properties of oxygen-precipitated silicon, Journal of Applied Physics 59 (1986) 2476.

54. D. Macdonald, The emergence of n-type silicon for solar cell manufacture, in: 50th Annual AuSES Conference (Solar 2012), Melbourne, Australia, 2012.

55. Cuevas, M.J. Kerr, C. Samundsett, F. Ferrazza, G. Coletti, Millisecond minority carrier lifetimes in n-type multicrystalline silicon, Applied Physics Letters 81 (2002) 4952.

56. T. Schutz-Kuchly, J. Veirman, S. Dubois, D.R. Heslinga, Light-InducedDegradation effects in boron–phosphorus compensated n-type Czochralski silicon, Applied Physics Letters 96 (2010) 093505.

57. M. Rinio, A. Yodyunyong, S. Keipert-Colberg, Y.P.B. Mouafi, D. Borchert, A. Montesdeoca-Santana, Improvement of multicrystalline silicon solar cells by a low temperature anneal after emitter diffusion, Progress in Photovoltaics: Research and Applications 19 (2011) 165.

# Chapter 8

# Challenges and prospects for developing CdS/CdTe substrate solar cells on Mo foils

**B.L. Williams [a,*], J.D. Major [a], L. Bowen [b], L. Phillips [a], G. Zoppi [c], I. Forbes [c], K. Durose [a]**

[a] Stephenson Institute for Renewable Energy, Liverpool University, Liverpool L69 7ZF, UK

[b] G. J. Russell Microscopy Facility, Durham University, South Road, Durham DH1 3LE, UK

[c] Northumbria Photovoltaics Applications Centre, Northumbria University, Ellison Building, Newcastle upon Tyne NE1 8ST, UK

## ABSTRACT

ITO/ZnO/CdS/CdTe/Mo solar cells have been grown in the substrate configuration by a combination of close-space sublimation and RF sputtering. A peak efficiency of 8.01% was achieved. A two stage $CdCl_2$ annealing process was developed, with the first stage contributing to CdTe doping and the second being linked to CdTe/CdS interdiffusion by secondary ion mass spectrometry analysis. The inclusion of a ZnO layer between CdS and ITO layers improved performance significantly (from $\eta$=6% to $\eta$=8%) by increasing the shunt resistance, *RSH*, from 563 Ω $cm^2$ to 881 Ω $cm^2$. Cross-sectional scanning electron microscopy highlighted the importance of the resistive ZnO layer as numerous pinholes and voids exist in the CdS film. Solar cell performance was also investigated as a function of CdTe thickness, with optimal thicknesses being in the range 3–6 μm. All devices were deemed to be limited principally by a non-Ohmic back contact, the Schottky barrier height being determined to be 0.51 eV by temperature dependent *J–V* measurements. Modelling of device performance using SCAPS predicted efficiencies as high as 11.3% may be obtainable upon formation of an Ohmic back-contact. SCAPS modelling also demonstrated that a quasi-Ohmic back-contact may be achievable via inclusion of a highly p-doped (~$10^{18}$ $cm^{-3}$) buffer layer, between CdTe and Mo, which also has an optimal electron affinity (4.2 eV). The evaluation of device processing and the in-depth characterisation presented here provides a number of insights towards the continued improvement of substrate cell performance.

## KEYWORDS

CdTe; Substrate; Thin-film; Optimisation; Interdiffusion; Back contact

## 1. INTRODUCTION

Recent significant increases in CdTe photovoltaic device performance [1] has followed a 20 year period of more modest progress [2]. The principal limitation that remains is considered to be the low open-circuit voltage, $VOC$ that is presently achievable (champion devices have $VOC$ in the range 850–900 mV) relative to CdTe's optical band gap (~1.5 eV): the inability to achieve sufficient doping densities (> $10^{16}$ $cm^{-3}$) due to self-compensation [3] and recombination at the CdS/CdTe interface [4] are likely explanations. Currently, the vast majority of CdTe solar cell research is based on the metal/CdTe/CdS/TCO/glass 'superstrate' configuration (where TCO is transparent conducting oxide) whereby device layers are deposited onto a glass superstrate [5],[6] and [7]. An alternative is the development of the TCO/CdS/CdTe/metal 'substrate' configuration (whereby the metal used is typically Mo due to it having a thermal expansion coefficient similar to that of CdTe). The substrate configuration offers two significant advantages: (a) the stages of CdTe doping and CdTe/CdS junction formation may be optimised independently and (b) the use of low cost, flexible metal substrates may enable roll-to-roll cell production and allow a wide range of applications, including building- and vehicle-integrated PV.

Research into substrate devices has been limited to date, with the best reported performances being shown in Table 1[8], [9], [10], [11], [12], [13] and [14]. The common finding for all substrate devices is that generating an Ohmic back contact is problematic. As noted in Table 1, this may be overcome by the inclusion of a 'buffer layer' between the Mo substrate and CdTe layer to improve the contact. For instance, the devices reported by Kranz et al. [8] and Dhere et al. [10], use Te/$MoO_3$ and CuTe buffer layers respectively. However, since the back contact is formed early in the process, it must not degrade throughout the remaining high temperature processes. In superstrate cells on the other hand, back contact formation is the final procedure of device fabrication and this is not an issue. Note that the 13.6% Kranz [8], 11.3% Gretener [9] and 11.0% Dhere [10] devices were grown on Mo/glass substrates: The best reported substrate CdTe device on a *metal*substrate is Kranz's 11.5% device on Mo foil [8].

Whereas most reports of substrate devices so far have tended to focus on the optimisation of performance rather than in-depth characterisation and modelling, here, all of these aspects are considered as we believe they are critical for identifying device limitations. First, we present optimisation studies of the post-growth device processing in order to obtain high efficiency substrate devices, with particular focus on the sequencing of the Cl-doping stage. Second, an extensive study of the effect of annealing temperature and absorber layer thickness on device performance is presented. In-depth characterisation of substrate devices is provided by focussed ion beam-scanning electron microscopy (FIB-SEM), external quantum efficiency (*EQE*) and current–voltage–temperature (*J–V–T*) measurements. Finally, device modelling, using SCAPS[15], is presented. We report a peak efficiency of 8.01% for ITO/ZnO/CdS/CdTe/Mo substrate cells (where ITO is indium tin oxide), grown by a combination of close-space sublimation and RF sputtering.

# 2. EXPERIMENTAL

## 2.1. Fabrication of ITO/CdS/CdTe/Mo substrate devices

The 50 mm×50 mm Mo foil substrates (Advent, 3 N pure) were cleaned ultrasonically in de-ionised water and dried with a $N_2$ air gun. CdTe films were close space sublimation-deposited, with a powdered source (Alfa Aesar, 5 N pure) heated to 605 °C and the substrate held at 530 °C. Growth was performed via a two stage process; firstly, under a static pressure of 25 Torr $N_2$ for 2–10 min, and then under low vacuum ($10^{-1}$ Torr) for 1 min. Growth at the higher pressure has been shown to nucleate larger grains [6], but since it also acts to slow the growth rate, the films were completed by growth at the lower pressure. This process generated films of thickness 1.8–10 μm. CdS window layers of thickness ~200 nm were RF sputtered under 5 mTorr Ar for 30 min, using a target power of 60 W, with the substrate held at 200 °C. The ITO front contact layer was also deposited by sputtering, under 1 mTorr Ar for 120 min, using a target power of 70 W and with the substrate at 350 °C. Whilst this completed the basic layer structure, in some samples an intrinsic ZnO film was included between the CdS and ITO layers, this also being grown by sputtering, at room temperature, under 1 mTorr Ar for 20 min, using a target power of 100 W. The 50 mm × 50 mm samples were then quartered into 25 mm×25 mm plates for device processing.

Preliminary growth runs indicated that the standard $CdCl_2$ annealing process used for superstrate devices (typically a 20–30 min anneal of the CdS/CdTe stack at 420 °C following $CdCl_2$ evaporation [16]) was ineffective for substrate cell activation. Hence, in order to optimise the device processing, four different $CdCl_2$ post-growth treatments were tested, these being summarised in Table 2 and now described:

Treatment A: After CdS deposition, 200 nm $CdCl_2$ was thermally evaporated and the structures were annealed in air at 420 °C for 30 min, with ITO being subsequently deposited, and no further treatment being applied.

Treatment B: $CdCl_2$ deposition and annealing conditions were identical to treatment A but were applied *prior* to CdS and ITO deposition, i.e. only the CdTe layer was treated.

Treatment C: As treatment B (i.e. after CdTe deposition) but with the addition of a *second*annealing step after ITO deposition, in air at 480–580 °C air for 30 min, with no additional $CdCl_2$ deposition.

Treatment D: No $CdCl_2$ deposition and just a single annealing step, after ITO deposition, in air at 480–580 °C for 30 min.

**Table 1.** Efficiency of CdTe devices grown in substrate configuration on Mo substrates. The 'buffer layer' referred to is situated between Mo and CdTe for all devices.

| ***Reference*** | ***Best efficiency* (%)** | ***Comments*** |
|---|---|---|
| [8] | (a) 13.6 (Mo/glass substrate), (b) 11.5 (Mo foil substrate) | Te/$MoO_3$ buffer layer + Cu doping |
| [9] | 11.3 (Mo/glass substrate) | Cu/Te/$MoO_3$ buffer layer |
| [10] | 11.0 (Mo/glass substrate) | $Cu_xTe$ buffer layer |
| [11] | 7.8 | ZnTe:N buffer layer |
| [12] | 6.5 | RF sputtered CdTe + unnamed buffer |
| [13] | 6.0 | ZnTe buffer layer |
| [14] | 6.0 | CuTe buffer layer |

**Table 2.** Annealing processes applied to substrate solar cells. Each annealing step was carried out in air for 30 min.

| *Treatment* | *After CdTe dep.* | *After CdS dep.* | *After ITO dep.* |
|---|---|---|---|
| A | – | 200 nm $CdCl_2$ $T_{CdCl2}$=420 °C | – |
| B | 200 nm $CdCl_2$ $T_{CdCl2}$=420 °C | – | – |
| C | 200 nm $CdCl_2$ $T_{CdCl2}$=420 °C | – | $T_{anneal}$=480–580 °C |
| D | – | – | $T_{anneal}$=480–580 °C |

For clarity, for the remainder of this paper, the temperature of the $CdCl_2$ annealing step after CdTe or CdS deposition is denoted as *$TCdCl_2$*, and the temperature of the annealing step after ITO deposition is denoted as *Tanneal*. After layer growth and annealing, devices were defined by mechanical scribing into 2.5 mm×2.5 mm squares. Device performance parameters quoted in Table 3 and Table 4 are from individual devices, whereas those plotted in Fig. 4 are averages taken from each plate (each having 16 individual devices).

**Table 3.** A comparison of the performance parameters of devices subjected to different post-growth annealing processes (see Table 2 and text).

| *Treatment* | $\eta$ (%) | $J_{SC}$ (mA/cm$^2$) | $V_{OC}$ (mV) | *FF* (%) | $R_S$ (Ω.cm$^2$) | $R_{SH}$ (Ω.cm$^2$) |
|---|---|---|---|---|---|---|
| A | 0.62 | 7.22 | 251 | 34.1 | 18.9 | 75 |
| B | 1.96 | 15.1 | 365 | 35.5 | 16.2 | 434 |
| C ($T_{2nd}$=560 °C) | 6.05 | 20.2 | 635 | 47.1 | 12.3 | 563 |
| D ($T_{2nd}$=560 °C) | 1.14 | 9.06 | 367 | 34.2 | 20.4 | 111 |

## 2.2. Characterisation and modelling

*J–V* measurements were carried out under AM1.5 illumination using an Oriel solar simulator, and *J–V–T* measurements were conducted in the dark, in the temperature range 200–350 K. *EQE* data was collected using a Bentham PVE 300 EQE system with white light bias. For secondary electron mass spectrometry (SIMS) analysis, a $O^{2+}$ ion gun (Hiden Analytical) was used to sputter the sample (using a beam energy of 5 keV) and secondary ions were analysed by a quadrupole detector. Cross-sectional SEM images were generated using a FIB-milling procedure [16] in a FEI Helios Nanolab 600: After C and Pt coating of the sample, a focussed beam of $Ga^+$ ions were used to mill a trench in the sample, and to polish a cross-section. This was then imaged in secondary electron mode in the same instrument. Modelling of solar cell *J*–*V* curves and electronic band-diagrams was carried out using SCAPS, v3.2.00 [15].

**Table 4**. The best contacts from the series of devices presented in Section 3.3, showing the effect of CdTe thickness and annealing temperature. For averaged data, see Fig. 4.

| $T_{anneal}$ (°C) | *Thickness* (μm) | $\eta$ (%) | $J_{SC}$ (mA/cm$^2$) | $V_{OC}$ (mV) | *FF* (%) |
|---|---|---|---|---|---|
| 540 | 1.8 | 0.49 | 5.63 | 219 | 39.8 |
| 540 | 3.4 | 0.74 | 9.35 | 261 | 30.4 |
| 540 | 6.1 | 3.01 | 18.4 | 478 | 34.2 |
| 540 | 10 | 2.08 | 10.5 | 560 | 35.4 |
| 560 | 1.8 | 2.08 | 15.9 | 339 | 38.5 |
| 560 | 3.4 | 2.98 | 19.8 | 400 | 37.6 |
| 560 | 6.1 | 4.29 | 18.0 | 541 | 44.1 |
| 560 | 10 | 2.49 | 9.25 | 622 | 43.4 |
| 580 | 1.8 | 5.30 | 21.6 | 581 | 42.3 |
| 580 | 3.4 | 5.08 | 22.1 | 578 | 39.7 |
| 580 | 6.1 | 4.55 | 18.9 | 581 | 41.5 |
| 580 | 10 | 4.29 | 20.4 | 601 | 35.0 |

# 3. RESULTS AND DISCUSSION

## 3.1. The impact of the $CdCl_2$ activation step

Fig. 1 shows the *J*–*V* curves of the best performing ITO/CdS/CdTe/Mo devices obtained from each of the post-growth treatments A–D, the shapes of the curves being representative of each respective sample set. Table 3 shows the photovoltaic performance parameters of these 'best devices'. All devices described in this comparison had a CdTe thickness, $X$=4±0.5 μm and did not include a ZnO layer. Those subject to treatments C and D were done so using *Tanneal*=560 °C.

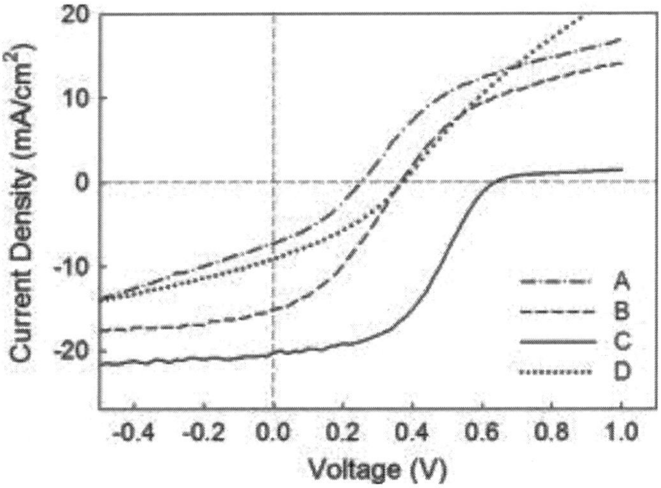


**Figure 1.** *J–V* curves of the best performing substrate devices obtained from each of the post-growth treatments A–D. See Table 2 and the text in Section 2 for a description of these treatments, and Table 3 for the associated solar cell performance parameters.

All devices subject to post-growth treatment A, i.e. annealing after CdS deposition and with no further annealing, performed very poorly, the majority of contacts were short-circuits while the remainder typically had $\eta$<0.5% due to low fill-factor, *FF* (<30%), short-circuit current, *JSC* (<5 mA/cm$^2$), and *VOC* (<300 mV). The best-device from this sample set had $\eta$=0.62%, limited by a series resistance (*RS*=18.9 Ω cm$^2$) that is considerably greater than that of high efficiency CdTe/CdS devices, and low *RSH* (75 Ω cm$^2$). Here, direct $CdCl_2$ treatment of the CdS could induce densification and pinholing, and hence the creation of shunting pathways. Moreover, Singh et al. [14] report that upon direct Cl-treatment of the CdS layer in substrate devices, there is an excessive intermixing at the CdS/CdTe interface resulting in the entire CdS layer being consumed and becoming $CdTe_xS_{1-x}$, lowering *VOC* to~0 V. Indeed, the presence of Cl is known to enhance S diffusion into CdTe [17].

Devices subject to post-growth treatment B, i.e. annealing after CdTe deposition and with no further annealing, also performed poorly, with the highest efficiency achieved being$\eta$=1.96%. Nevertheless, relative to treatment A, these devices had slightly improved *VOC*(in the range 200–400 mV), and *JSC* (5–15 mA/cm$^2$), whilst *FF* remained<35%. Since treatment B yielded slightly higher *RSH* (434 Ω $cm^2$ for the best device,~400 Ω $cm^2$ being typical for the sample set) than treatment A, it appears as though direct $CdCl_2$-treatment of CdTe is more effective than treatment of CdS/CdTe stacks in terms of avoiding pinholing. Nevertheless *RSH* remains much lower than that of high efficiency devices, and *RS* is too high (16.2 Ω $cm^2$ for the best device,~20 Ω $cm^2$ being typical for the sample set).

Post-growth treatment C yielded significant performance enhancements, this differing from treatment B only by the inclusion of a second annealing stage following ITO/CdS deposition. Whereas the peak efficiency for treatment B was $\eta$=1.96%, a device subject to treatment C with *Tanneal*=560 °C had $\eta$=6.05%. This was principally a result of enhancements to *VOC* (from 365 mV for treatment B to 635 mV for treatment C) and *FF*(from 35.5% to 47.1%), implying an improvement in junction quality, this being discussed below with support from compositional data. For treatment C, *RS* (12.3 Ω $cm^2$) was lower than for either treatments A or B, and *RSH* (563 Ω $cm^2$) was higher.

All devices that were subject to treatment D, which consisted of a single annealing stage after ITO deposition but with no prior Cl-doping, had efficiencies, $\eta$<1.2%. The low *VOC*(367 mV for the best device) is to be expected for devices that are not $CdCl_2$-treated since Cl-doping is typically necessary in order to attain sufficiently high p-type doping densities.

For all devices, there is a strong roll-over of the *J–V* curve in the first quadrant, which indicates a significantly poor back-contact. This was expected owing to the low work function of Mo (~4.36–4.95 eV) [18] relative to CdTe (> 5.7 eV) and the absence of a back contact buffer layer. The effect of the back contact on device performance is investigated in more detail in Section 3.6.

From these results, it is clear that the most effective post-growth treatment was that of using a twice-annealing process (treatment C). Whilst the first annealing stage, i.e. Cl-activation of the CdTe layer, is presumed to p-type dope the CdTe, the impact of the second annealing stage on the composition in the devices was investigated using SIMS profiling. Two devices were analysed, one subject to treatment B (having $\eta$=1.96%) and one subject to treatment C (having $\eta$=6.05%). Figs. 2a and b shows the S and Cl distributions respectively as a function of etch time for both devices. The left edge of the plots represents the front surface of the cell, i.e. the ITO surface. We do not quote the actual depth that the etch time correlates to, but we infer that the depth of the CdS layer is represented by the S peak of the single-annealed sample (B). It is clear from the data that S diffusion into the CdTe layer is enhanced by the second annealing stage. Whilst excessive interdiffusion is detrimental to device performance (causing CdS consumption), a certain level of interdiffusion is thought to assist in the passivation of surface defects at the interface and therefore lead to reduced interface recombination. The improvements in *VOC* and *JSC* affected by the second annealing step are consistent with this. There is also a distinct difference in the Cl-profiles of the two devices. In the device subject to just one annealing stage (B), there is an excess of Cl deep in the CdTe layer, towards the CdTe/Mo interface. In the device subject to two annealing stages (C), the deep Cl peak is not seen but there is a higher Cl concentration in the CdS layer and throughout the remainder of the CdTe, presumably having diffused from deep in the CdTe following the second-annealing stage. This is likely to induce morphological changes in the junction region.

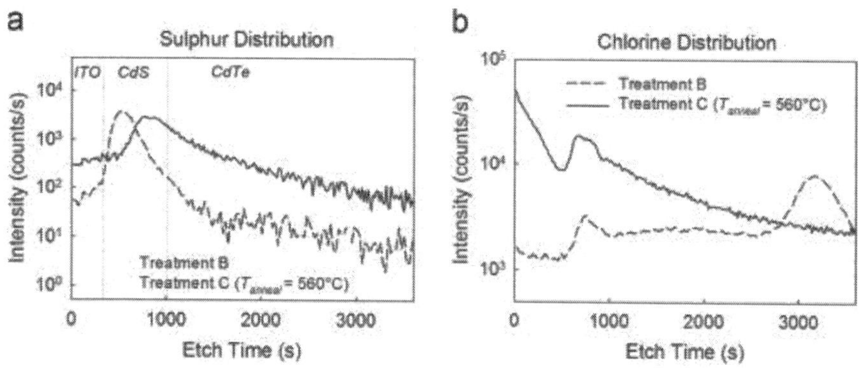


**Figure 2.** SIMS depth-profiles of (a) sulphur and (b) chlorine concentrations in a device subject to treatment B (single-anneal) and a device subject to treatment C (double-anneal, *Tanneal*=560 °C). The left edge of the plots represents the front surface of the cell, i.e. the ITO surface.

## 3.2. Inclusion of ZnO layer and EQE of best device

Fig. 3a shows the *J–V* response of devices subject to treatment C at *Tanneal*=560 °C with and without the inclusion of a ZnO layer between the CdS and ITO layers. The device without ZnO was the device having *η*=6.05% that was presented in the previous section.

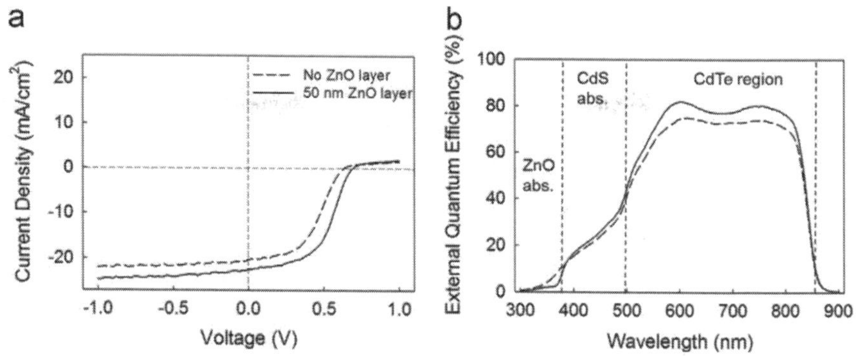


**Figure 3.** (a) *J–V* curves and (b) *EQE* response of devices subject to treatment C at *Tanneal*=560 °C with and without the inclusion of a ZnO layer.

Upon inclusion of the ZnO layer, device performance is clearly enhanced, the efficiency increasing to $\eta$=8.01%. This resulted from improvements to *JSC* (from 20.2 mA/cm$^2$ to 22.4 mA/cm$^2$), *VOC* (from 635 mV to 690 mV) and *FF* ( from 47.1% to 51.9%). Notably*RSH* is significantly improved upon inclusion of the ZnO layer, from 563 Ω cm$^2$ to 881 Ω cm$^2$. This device had the best $\eta$ achieved in these experiments. The mechanism of improvement upon use of such a high resistive layer is thought to be that it limits performance losses by electrically 'blocking' shunting paths and weak ITO/CdTe diodes (caused by pinholes), therefore increasing *RSH* and *VOC*, this being consistent with the results presented here.

Fig. 3b shows the *EQE* response measured from this device. The main features of the response from the device including a ZnO layer are now described. A relatively flat response is observed in the spectral region 600–860 nm, peaking at~80% – the drop at~860 nm being at the CdTe band-edge. There are significant losses in the above-gap region of the CdS (i.e. ≤500 nm) and a further reduction in *EQE* for $\lambda$<390 nm associated with absorption by the ZnO. Note also the gradual decrease in *EQE* in the 500–550 nm wavelength region, presumably associated to absorption in alloyed CdS1-yTey for which the band-gap is lower than CdS. A similar response was observed for the device without ZnO, the only differences being (a) a lower efficiency in the CdTe region and (b) the absence of the ZnO related absorption cut-off at 390 nm.

## 3.3. Effect of CdTe thickness

A series of devices with CdTe thickness, *X*, in the range 1.8–10 µm were grown, subject to treatment C at *Tanneal*=540–580 °C. For all these devices, a 50 nm ZnO film was included between the CdS and ITO layers. The average performance parameters as a function of *X* and *Tanneal* are plotted in Fig. 4. The parameters of the *best* contacts from each plate are listed in Table 4 for reference, although the discussion that follows generally refers to the averaged data. In Fig. 5, typical *J–V* curves for devices subject to treatment C with *Tanneal*=580 °C, having different *X* are shown. Note that this was an entirely new sample set and the 'best' device reported in the previous section was not part of this set.

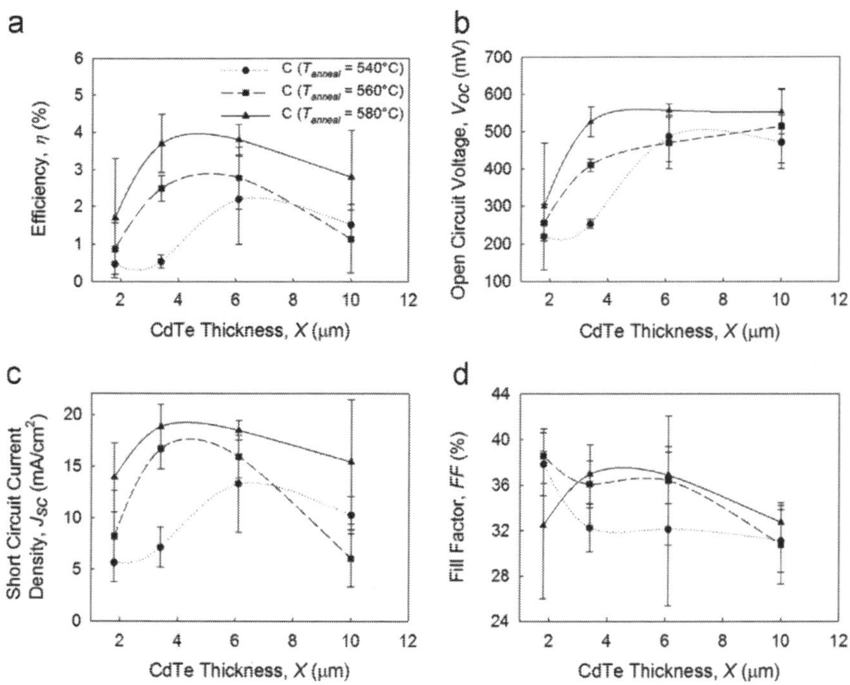


**Figure 4.** Dependence of (a) *η*, (b) *VOC*, (c) *JSC* and (d) *FF* as a function of CdTe thickness for a range of annealing temperatures (*Tanneal*=540–580 °C). Lines added as a guide to the eye. The data points show the averages and standard deviation for up to 16 devices. See Table 4 for the best contacts from each plate.

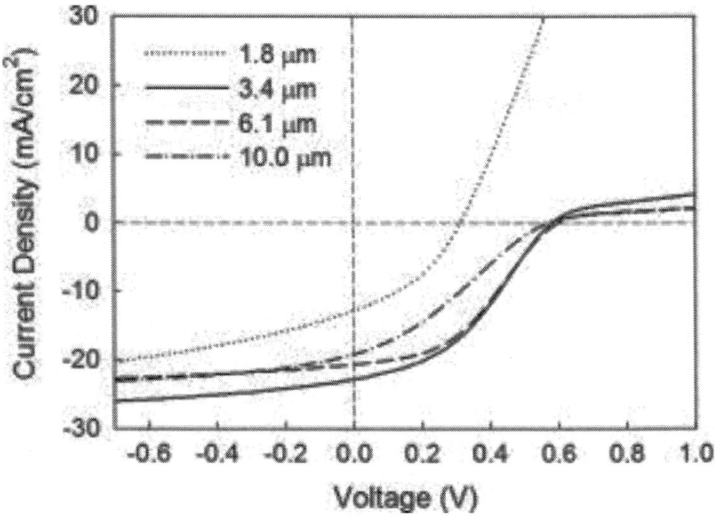


**Figure 5.** Typical *J–V* curves for devices subject to treatment C (*Tanneal*=580 °C), with different CdTe thicknesses (1.8–10 µm).

Regardless of annealing temperature the optimum CdTe thickness was *X*=3–6 µm, the average *η* being ~4% in this range for *Tanneal*=580 °C. The thickness-dependency of *η*closely followed that of *JSC*, which peaked at 18.9±2.1 mA/cm$^2$ with an absorber thickness of 3.4 µm, also for *Tanneal*=580 °C.

As well as having lower average *JSC* (<14 mA/cm$^2$), the thinnest devices (*X*=1.8 µm) typically suffered from much lower average *VOC* (<300 mV) than the rest of the samples resulting in average efficiencies of ≤1.8% for all annealing temperatures. This may directly result from the presence of micro-shorts in the thinner CdTe layers, as suggested by Clayton et al. [19], although this should also result in a reduction of *FF* which is not consistently observed here. Instead, it may be attributed to the Schottky back contact at the Mo/CdTe interface: devices with a significant back contact barrier are usually modelled to consist of two independent and opposing diodes [20] (whereby the back contact diode acts to increase *RS*), but in the case of thin absorber layers the depletion region of the back contact will overlap with that of the main junction and therefore the *VOC*will be reduced. In accordance with the failure of the independent two-diode model, no rollover should be observed, as is the case here for the thinnest device (1.8 µm) in Fig. 5. The performance from these thinnest devices was notably inconsistent, with one device having an efficiency as high as $\eta$=5.3% (see Table 4).

The highest *VOC*s (approaching 600 mV) were achieved when using the highest annealing temperature, implying greater levels of S and Te intermixing and/or recrystallisation of CdTe/CdS layers. For all annealing temperatures, the *VOC* plateaus above a CdTe thickness of 6 µm.

The thickest devices (10 µm) suffered from lower *FF* (30–35%) for all annealing temperatures. Indeed, for the series of devices annealed at *Tanneal*=580 °C, *RS* increased with CdTe thickness: *RS* was calculated from the slope of the typical *J–V* curves ( Fig. 5) in forward bias to be 8.1 Ω $cm^2$ (for CdTe thickness of 1.8 µm), 14.1 Ω $cm^2$ (3.4 µm), 14.8 Ω $cm^2$ (6.1 µm) and 22.3 Ω $cm^2$ (10.0 µm). This may be due to the higher probability that there exist grain boundaries perpendicular to current flow in thicker CdTe films, or simply due to the intrinsic resistance of the CdTe itself. Notably, however, when using a lower annealing temperature of 540 °C, the *RS* for thinner CdTe films (3–6 µm) was equivalent to that for the 10 µm CdTe films that had been annealed at 580 °C, this implying that high *RS* is attributed to CdTe films being 'undertreated'.

## 3.4. FIB-SEM

Fig. 6 shows secondary electron SEM images of a FIB-milled cross-section from a ITO/ZnO/CdS/CdTe/Mo device. This device was subject to treatment C (*Tanneal*=580 °C), having a CdTe thickness of~1.8 µm and efficiency of 5.3% (taken from the sample set presented in Section 3.2). The grains in the CdTe layer are relatively large, extending the entire width of the film and up to 4 µm in width. There are low-energy twin boundaries present, but these are considered to be non-electrically active [21]. Most notable is the existence of voids (diameter ~100 nm) both in the bulk of the CdTe and in the CdS. Moreover, the higher magnification image (Fig. 6b) shows that there are discontinuities (or pinholes) in the CdS film, resulting in the ZnO coming into contact with the CdTe. This further emphasises the importance of using a ZnO layer if the CdS uniformity is insufficient, as it minimises the impact of weak diode regions (e.g. ITO/CdTe) which are detrimental to *VOC*. It was also observed that the CdS layer is significantly thinner (~120 nm) after the second annealing step than directly after growth (~200 nm). This further supports the SIMS data that S diffusion into the CdTe occurs during annealing.

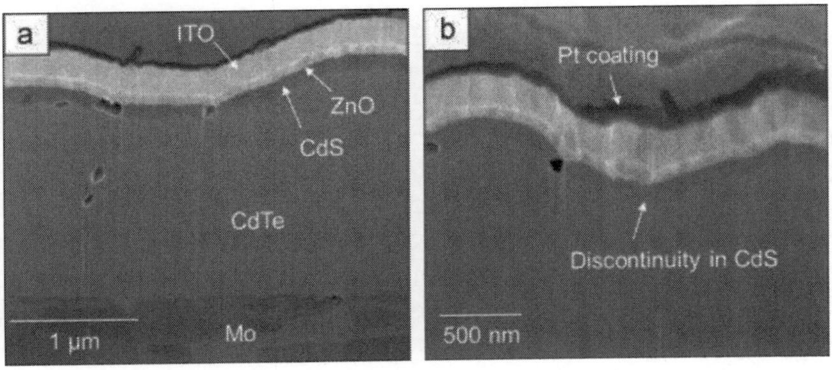


**Figure 6.** Secondary electron images of FIB-milled ITO/ZnO/CdS/CdTe/Mo device cross-section: (a) Full device structure and (b) the junction region. The device was subject to treatment C, i.e. the double-anneal process, with $T_2nd$=580 °C, and had efficiency, $\eta$=5.3%.

## 3.5. *J–V–T* measurements

*J–V–T* measurements were used to determine the back contact barrier height of a device subject to treatment C at *Tanneal*=580 °C and having a CdTe thickness of ~6 µm and$\eta$=5.5% (taken from a new sample set). The barrier height, $\varphi_B$, may be extracted from the temperature dependence of *RS* via the method of Bätzner et al. [22], which assumes that*RS* can be expressed as a summation of an Ohmic and an exponential component. Here, at each temperature, *RS* was calculated from the slope in the forward bias region of dark*J–V* curves. Fig. 7a shows the experimental *RS–T* data for this device, fitted with Bätzner's model, which yields a barrier height of $\phi_B$= 0.51 eV. Demtsu et al. [20] state that in the two-diode model of a photovoltaic device any barrier height that exceeds 0.5 eV is detrimental to the *FF*, although not necessarily to *VOC*.

Furthermore, by investigating the temperature dependence of the ideality factor, *n*, and saturation current, *J0*, information about the charge transport mechanisms operating within a device may be yielded, as reviewed extensively by Alturkestani [23]. At each temperature, *n* and *J0* may be extracted from the gradient and *x*-axis intercept of dark ln(*J*)–*V* plots respectively, provided the following assumption to the Shockley equation is made:

$$J = J_0\left[\exp\left(\frac{qV}{nk_BT}\right) - 1\right] \approx J_0 \exp\left(\frac{qV}{nk_BT}\right) \quad (1)$$

where *q* is the electron charge and *kB* is Boltzmann's constant. Fig. 7b shows the temperature dependence of the ideality factor, *n*, and saturation current, *J0* for the same device as above (subject to treatment C, *X*=6.1 μm, *η*=5.5%). In the temperature range 270–350 K, *n*~2 which indicates that transport occurs via recombination in the depletion region [24]. The ln(*J0*) ∝−1/*T* dependence is also indicative of a recombination regime in this temperature range [24]. However, below 270 K, *n*increases as the temperature is reduced, which indicates that charge transport is dominated by a multi-step tunnelling regime [25].

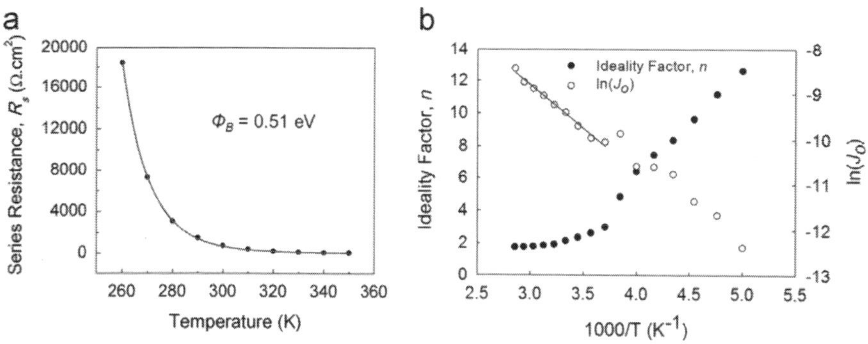


**Figure 7.** (a) The temperature dependence of *RS*, used to determine the back contact barrier height, $\phi_B$. (b) Temperature dependence of the ideality factor, *n*, and the saturation current, *J0*. This data was taken from a device subject to treatment C (*Tanneal*=580 °C), having a CdTe thickness of ~6 μm and for which *η*=5.5%.

## 3.6. SCAPS modelling and use of back contact buffer layers.

Device modelling was carried out using the SCAPS simulation tool developed by Burgelman et al. [15] in order to verify the *J–V* characteristics of the best performing device of this study and to determine to what extent the poor back contact limits device efficiency. The program enables the definition of the film stack, the assignment of material properties (e.g.

optical band gap, doping density and type, density of states) to each film and the inclusion of defect states in the bulk of films and at interfaces. For the back contact the only variables are the work function and barrier height. To simulate the devices reported here the material properties used by Gloeckler et al. [26] in a SCAPS simulation of a CdTe/CdS superstrate cell ($\eta$=16.4%) were used as a starting point (these include band gaps, optical properties, trap parameters) and then the model was modified to include (a) a back contact with a work function of~5.25 eV to represent Mo; (b) a CdTe electron affinity, *Eea*, of 4.5 eV; (c) a maximum CdTe doping density of $NA=2\times10^{15}$ $cm^{-3}$, as determined from preliminary capacitance–voltage measurements (not shown); (d) reduced *NA* in the CdTe layer towards the back contact due to the depletion of holes affected by the Schottky barrier; and (e) increased *RS*, comparable to devices reported here (10 Ω $cm^2$). *J–V* curves were simulated under AM1.5 illumination the device, and this was repeated once the option for 'Ohmic contact' was selected, which generates an Ohmic contact regardless of the properties of metal and CdTe.

Fig. 8 shows the experimental *J–V* curve for the best performing device, the simulated *J–V* curve that represents this device, and a simulated curve of the device with an Ohmic contact. A good agreement between the experimentally determined performance parameters ($\eta$=8.01%, *JSC*=22.4 mA/$cm^2$, *VOC*=690 mV, *FF*=51.9%) and the simulated parameters ($\eta$=8.02%, *JSC*=22.4 mA/$cm^2$, *VOC*=694 mV, *FF*=51.6%) was achieved. Upon implementation of an Ohmic contact, the simulation suggests these may be increased to $\eta$=11.3%, *JSC*=22.9 mA/$cm^2$, *VOC*=792 mV and *FF*=62.3%, with the curve showing no rollover. This demonstrates that the back contacting to our device is significantly hindering device performance and that the efficiency shortcoming of our devices relative to those of Kranz et al. [8] is largely attributable to the poor back contact.

The SCAPS model was also used to assess the impact of including various types of back contact buffer layer (with high p-type doping and/or optimised electronic band structure), between Mo and CdTe. A buffer layer of 500 nm thickness was added to the device model between CdTe and Mo, with all properties being kept the same as that of CdTe (e.g. optical band-gap, dielectric permittivity, densities of states, electron and hole mobilities), but with the value of its electron affinity, *Eea*, or p-type doping density, *NA*, being modified. The chosen values of the buffer layers' doping density and electron affinity are listed inTable 5, along with the simulated performance parameters of the devices upon the buffer layers' inclusion. The electronic band diagrams at the Mo/buffer layer/CdTe interfaces, also generated by SCAPS (at zero bias in the dark), for each of the simulated buffer layers are shown in Fig. 9.

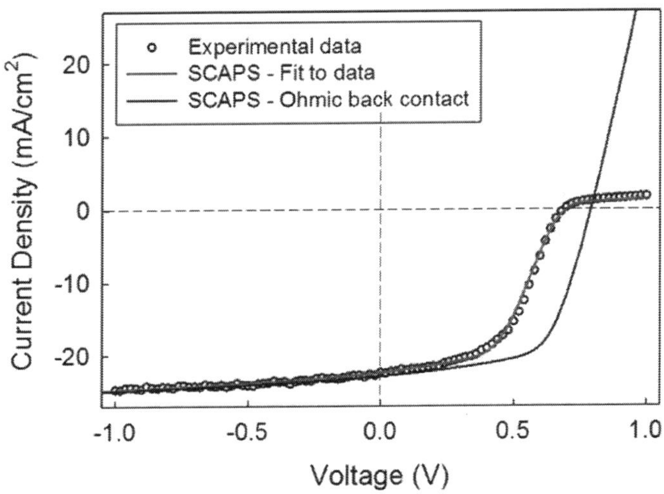


**Figure 8.** *J–V* data for highest efficiency ($\eta$=8.01%) substrate cell overlapped with modelled *J–V* data from SCAPS software ($\eta$=8.02%). By inclusion of an Ohmic back contact in the model, the predicted performance was improved to 11.3%.

**Table 5.** Performance characteristics of substrate configuration devices given by SCAPS software when various back contact buffer layers are added to the original structure.

| *Buffer layer* $N_A$ | *Buffer layer* $E_{ea}$ | *Buffer thickness* | $\eta$ (%) | $V_{OC}$ (mV) | $J_{SC}$ (mA/cm$^2$) | *FF* (%) |
|---|---|---|---|---|---|---|
| No buffer (CdTe) | | – | 8.02 | 694 | 22.4 | 51.6 |
| $10^{15}$ cm$^{-3}$ | 4.2 eV | 500 nm | 10.61 | 791 | 22.4 | 59.9 |
| $10^{18}$ cm$^{-3}$ | 4.5 eV | 500 nm | 9.65 | 772 | 22.3 | 56.1 |
| $10^{18}$ cm$^{-3}$ | 4.2 eV | 500 nm | 11.32 | 794 | 22.5 | 63.4 |

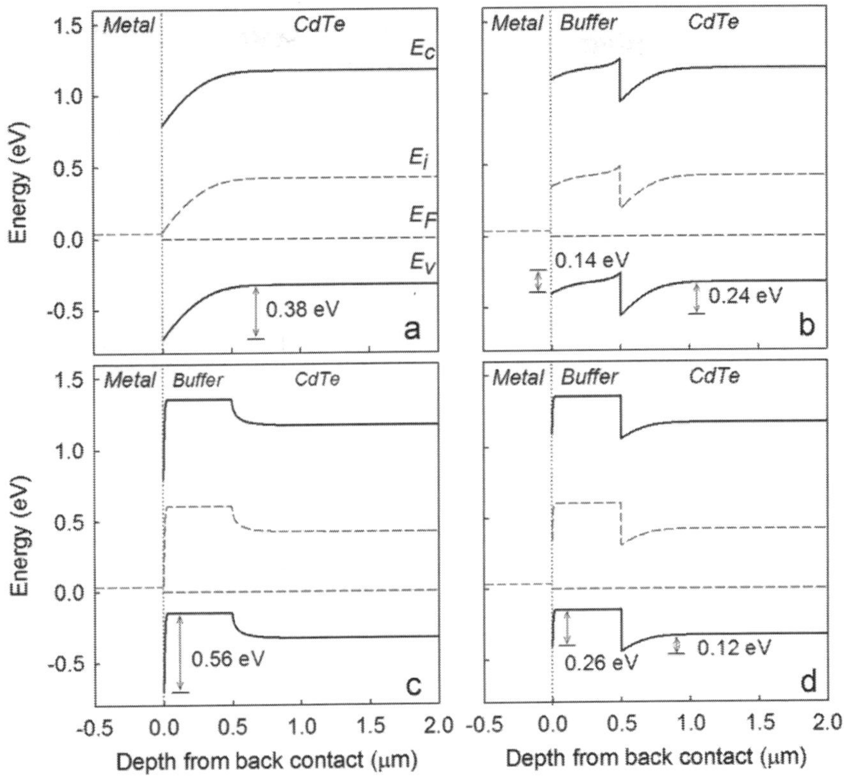


**Figure 9.** Electronic band diagrams (at zero bias in the dark) generated using SCAPS focussing on the back contact region and the effect of the use of various buffer layers, each having a thickness of 500 nm. (a) No buffer layer, (b) $NA=10^{15}$ $cm^{-3}$ and $Eea=4.2$ eV, (c) $NA=10^{18}$ $cm^{-3}$ and $Eea=4.5$ eV, (d) $NA=10^{18}$ $cm^{-3}$ and$Eea=4.2$ eV.

With no buffer layer present the modelling was exactly the same as for the fit to the experimental data in Fig. 8 for which the efficiency was 8%. The band diagram for this device (Fig. 9a) shows there is a barrier height of 0.38 eV, this being slightly lower than the experimental value (presumably due to the simulation being carried out at zero bias).

By choosing a buffer layer with an electron affinity (4.2 eV) such that it's work function, $\psi$,lies between the values of the metal and CdTe (i.e. 5.25 eV<$\psi$<5.9 eV), the barrier at the back contact can be split into two smaller barriers (Fig. 9b). The effect of this was to significantly improve the *VOC* (from 690 mV to 790 mV) and consequently $\eta$ (from 8.0% to 10.6%).

Alternatively, by choosing a highly p-doped ($10^{18}$ $cm^{-3}$) buffer layer, the barrier at the metal/buffer interface is large but narrow (Fig. 9c), allowing tunnelling to occur, and an improvement in $\eta$ (to 9.7%) is achieved. Indeed, this approach is used for the formation of Ohmic contacts to superstrate CdTe devices: the CdTe back surface is usually chemically etched to generate a Te-rich surface making it highly p-type. However, for substrate cells, etching of the back surface is not possible. To yield a performance as high as that achieved with an Ohmic contact (11.3%, represented by the solid *J–V* curve in Fig. 8), a buffer layer that is both highly doped and has an ideal electron affinity should be included (Fig. 9d), this generating one small barrier at the buffer layer/CdTe interface (0.12 eV) and one narrow barrier at the metal/buffer layer interface (0.26 eV). These results demonstrate what properties a buffer layer must possess, and they imply that the shortcoming of our devices is dominated by the non-Ohmic back contact. Fahrenbruch[27] has reviewed similar methods to those proposed here, citing $Cu_xTe$, $Sb_2Te_3$ and ZnTe:N as potential buffer layer materials with sufficiently high $N_A$ (~$10^{21}$,~$10^{20}$and~$10^{18}$ $cm^{-3}$ respectively) but no materials with $\psi_{metal}<\psi<\psi_{CdTe}$.

## 4. CONCLUSIONS

A peak efficiency of 8.01% has been achieved for substrate configuration CdTe solar cells, with the device architecture ITO/ZnO/CdS/CdTe/Mo.

Initially, the effect of varying the sequence of the $CdCl_2$ activation step was investigated. It was shown to be optimal to use a double-annealing process; (a) $CdCl_2$ deposition onto the CdTe layer acts to dope the CdTe and (b) high temperature (> 560 °C) annealing following CdS and ITO deposition improves the junction quality by enhancing S and Te interdifussion. Increased S diffusion into the CdTe layer following the second annealing stage was demonstrated using SIMS analysis and supported by the observation that the CdS layer is thinner following the second anneal than immediately after growth. Other post-growth treatments that involved just a single-annealing step resulted in generally low efficiencies ($\eta$<2%).

Improvements to device efficiency were made by including a ZnO layer between the CdS and ITO. These improvements were mainly realised through increased $V_{OC}$ and $R_{SH}$therefore it is presumed that the ZnO layer limits the detrimental effects caused by pinholes. Indeed, voids and pinholes in the CdS layer were observed in FIB-SEM images.

The effect of the thickness of CdTe on device performance was also studied, with the optimum thickness being in the range 3–6 μm. Devices with thinner CdTe (1.8 μm) suffered from poor *VOC*, whereas those with thicker CdTe layers (10 μm) had low *FF* due to increased *RS*.

The *J–V* curves of all devices studied here exhibited significant roll-over, this being attributed to a non-Ohmic back contact, the barrier height being determined to be as high as 0.51 eV from *J–V–T* measurements. *J–V–T* measurements also indicate that in the temperature range 270–350 K, charge transport is dominated by recombination in the depletion region, whereas at lower temperature a multi-step tunnelling regime operates.

SCAPS modelling implied that the non-Ohmic contact was the principal limiting factor to efficiency for these devices, and that the successful formation of an Ohmic contact would yield efficiencies equivalent to the highest reported for substrate cells. A quasi-Ohmic contact may be achieved through inclusion of a buffer layer (at the Mo/CdTe interface) which is highly p-doped ($10^{18}$ $cm^{-3}$) and has an electron affinity~4.2 eV.

## ACKNOWLEDGEMENTS

The authors acknowledge EPSRC funding, and also Prof. M. Burgelman for the open access to the SCAPS software.

## REFERENCES

1. First Solar Inc. Press Release ⟨http://investor.firstsolar.com/Releasedetail.Cfm? Releaseid=743398⟩ (accessed on 14/11/13).

2. J. Britt, C. Ferekides, Thin-film CdS/CdTe solar cell with 15.8% efficiency, Appl. Phys. Lett. 62 (1993) 2851–2852.

3. A.W. Brinkman, CdTe and related compounds, in: R.T. Triboulet, P. Siffert (Eds.), Physics, Defects, Hetero- and Nano-Structures, Crystal Growth, Surfaces and Applications, Elsevier, Amsterdam, 2010.

4. J.R. Sites, J.E. Granata, J.F. Hiltner, Losses due to polycrystallinity in thin-film solar cells, Sol. Energy Mater. Sol. Cells 55 (1998) 43–50.

5. X.Z. Wu, High-efficiency polycrystalline CdTe thin-film solar cells, Sol. Energy 77 (2004) 803–814.

6. J.D. Major, Y.Y. Proskuryakov, K. Durose, G. Zoppi, I. Forbes, Control of grain size in sublimation-grown CdTe, and the improvement in performance of devices with systematically increased grain size, Sol. Energy Mater. Sol. Cells 94 (2010) 1107–1112.

7. Bosio, N. Romeo, S. Mazzamuto, V. Canevari, Polycrystalline CdTe thin films for photovoltaic applications, Prog. Cryst. Growth Charact. Mater. 52 (2006) 247–279.

8. L. Kranz, C. Gretener, J. Perrenoud, R. Schmitt, F. Pianezzi, F. La Mattina, P. Blosch, E. Cheah, A. Chirila, C.M. Fella, H. Hagendorfer, T. Jager, S. Nishiwaki, A.R. Uhl, S. Buecheler, A.N. Tiwari, Doping of polycrystalline CdTe for highefficiency solar cells on flexible metal foil, Nat. Commun. 4 (2013) 2306.

9. Gretener, J. Perrenoud, L. Kranz, L. Kneer, R. Schmitt, S. Buecheler, A.N. Tiwari, CdTe/CdS thin film solar cells grown in substrate configuration, Prog. Photovolt.: Res. Appl. 21 (2013) 1580–1586.

10. R.G. Dhere, J.N. Duenow, C.M. DeHart, J.V. Li, D. Kuciauska, T.A. Gessert, Development of substrate structure CdTe photovoltaic devices with performance exceeding 10%, in: Conference Record of the 38th IEEE PV Specialists Conference, Austin, TX, 2012.

11. Matulionis, S. Han, J. Drayton, K. Price, A. Compaan, Cadmium telluride solar cells on molybdenum substrates, MRS Proc. 668 (2001) 10.1557/PROC-668-H8.23.

12. J. Drayton, A. Vasko, A. Gupta, A.D. Compaan, Magnetron sputtered CdTe solar cells on flexible substrates, in: Conference Record of the31st IEEE Photovoltaic Specialists Conference, IEEE, New York, 2005, p. 406–409.

13. D.R. Hodges, V. Palekis, S. Bhandaru, K. Singh, D. Morel, E.K. Stefanakos, C.S. Ferekides, Mechanical properties and adhesion of CdTe/CdS thin film solar cells deposited on flexible foil substrates, in: A. Yamada, et al., (Eds.), Thin-Film Compound Semiconductor Photovoltaics, 1165, Materials Research Society, Warrendale, 2010, pp. 67–72.

14. V.P. Singh, J.C. McClure, Design issues in the fabrication of CdS/CdTe solar cells on molybdenum foil substrates, Sol. Energy Mater. Sol. Cells 76 (2003) 369–385. [15] M. Burgelman, P. Nollet, S. Degrave, Modelling polycrystalline semiconductor solar cells, Thin Solid Films 361 (2000) 527–532.

15. J.D. Major, L. Bowen, K. Durose, Focussed ion beam and field emission gunscanning electron microscopy for the investigation of voiding and interface phenomena in thin-film solar cells, Prog. Photovolt. 20 (2012) 892–898.

16. M. Terheggen, H. Heinrich, G. Kostorz, A. Romeo, D. Baetzner, A.N. Tiwari, A. Bosio, N. Romeo, Structural and chemical interface characterization of CdTe solar cells by transmission electron microscopy, Thin Solid Films 431 (2003) 262–266.

17. D.R. Lide, CRC Handbook of Chemistry and Physics, 89th Edition, CRC Press, Boca Raton, Florida, USA, 2008.

18. A.J. Clayton, S.J.C. Irvine, E.W. Jones, G. Kartopu, V. Barrioz, W.S.M. Brooks, MOCVD of Cd1-xZnxS/CdTe PV cells using an ultra-thin absorber layer, Sol. Energy Mater. Sol. Cells 101 (2012) 68–72.

19. S.H. Demtsu, J.R. Sites, Effect of back-contact barrier on thin-film CdTe solar cells, Thin Solid Films 510 (2006) 320–324.

20. K. Durose, G.J. Russell, Twinning in CdTe, J. Cryst. Growth 101 (1990) 246–250.

21. D.L. Batzner, M.E. Oszan, D. Bonnet, K. Bucher, Device analysis methods for physical cell parameters of CdTe/Cds solar cells, Thin Solid Films 361 (2000) 288–292.

22. M. Alturkestani, Cells: Key Layers and Electrical Effects (Ph.D. Thesis), University of Durham, UK, 2010.

23. S.M. Sze, Physics of Semiconductor Devices, 2nd Edition, Wiley, New Jersey, 1982.

24. S.S. Ou, O.M. Stafsudd, B.M. Basol, Current transport mechanisms of electrochemically deposited CdS/CdTe heterojunction, Solid-State Electron. 27 (1984) 21–25.

25. M. Gloeckler, A.L. Fahrenbruch, J.R. Sites, Numerical modeling of CIGS and CdTe solar cells: setting the baseline, in: Proceedings of 3rd World Conference on PV Energy Conversion, vols. A–C, WCPEC-3 Organizing Committee, Tokyo, 2003.

26. A.L. Fahrenbruch, Exploring back contact technology to increase CdS/CdTe solar cell efficiency, MRS Proc. 1012 (2007) 283–290, http://dx.doi.org/ 10.1557/PROC-1012-Y07-05.

# Chapter 9

# Highly transparent front electrodes with metal fingers for p-i-n thin-film silicon solar cells

**Etienne Moulin[1,2a,b,c], Thomas Christian Mathias Müller[2c], Marek Warzecha[2c], Andre Hoffmann[2], Ulrich Wilhelm Paetzold[2] and Urs Aeberhard[2c]**

[1] Ecole Polytechnique Fédérale de Lausanne (EPFL), Institute of Microengineering (IMT), Photovoltaics and Thin-Film Electronics Laboratory, Rue de la Maladière 71b, 2000 Neuchâtel, Switzerland

[2] IEK5-Photovoltaik, Forschungszentrum Jülich GmbH, 52425 Jülich, Germany

## ABSTRACT

The optical and electrical properties of transparent conductive oxides (TCOs), traditionally used in thin-film silicon (TF-Si) solar cells as front-electrode materials, are interlinked, such that an increase in TCO transparency is generally achieved at the cost of reduced lateral conductance. Combining a highly transparent TCO front electrode of moderate conductance with metal fingers to support charge collection is a well-established technique in wafer-based technologies or for TF-Si solar cells in the substrate (n-i-p) configuration. Here, we extend this concept to TF-Si solar cells in the superstrate (p-i-n) configuration. The metal fingers are used in conjunction with a millimeter-scale textured foil, attached to the glass superstrate, which provides an antireflective and retroreflective effect; the latter effect mitigates the shadowing losses induced by the metal fingers. As a result, a substantial increase in power conversion efficiency, from 8.7% to 9.1%, is achieved for 1-$\mu$m-thick microcrystalline silicon solar cells deposited on a highly transparent thermally treated aluminum-doped zinc oxide layer combined with silver fingers, compared to cells deposited on a state-of-the-art zinc oxide layer.

# 1 INTRODUCTION

Optimized transparent conductive oxide (TCO) layers applied as front electrodes in thin-film silicon (TF-Si) solar cells should address two trade-offs in order to enable cells to perform well both optically and electrically. The first trade-off is related to the surface morphology of the TCO layers. To efficiently absorb light in the optically thin silicon (Si) layers, and consequently achieve high photocurrent values, an advanced light-trapping concept has to be implemented. Light trapping in TF-Si solar cells is conventionally realized by applying randomly textured TCO electrodes to scatter light and thereby elongate the light path within the absorber layers [1, 2, 3, 4, 5, 6]; the compromise here resides in the fact that the TCO surface morphologies providing the most suitable light trapping in TF-Si solar cells often trigger the formation of defective porous regions in the subsequently grown Si layers, and these porous regions are detrimental for the electrical performance of the cells [7, 8].

The second trade-off is related to the intrinsic properties of the applied TCO layers. TCOs should exhibit a high conductivity and a high transparency in the full spectral range of relevance for solar cells, i.e. between 300 nm and 1100 nm. As the optical and electrical properties of TCOs are interlinked via the charge-carrier density and mobility, a high transparency in the near infrared (NIR) range is generally associated with a high sheet resistance $R_{sq}$ and vice versa [9, 10]. To compare the suitability of different TCOs to be used as electrode materials, a commonly accepted figure of merit (FOM) is defined as $FOM = \sigma/\alpha$, where $\sigma$ is the film conductivity in S cm$^{-1}$ and $\alpha$ is its absorption coefficient in cm$^{-1}$ at a given wavelength [11], the best TCO being the one with the highest conductivity and the lowest light absorption.

The most common TCOs used as front electrodes for TF-Si solar cells in the superstrate (p-i-n) configuration are aluminum-doped zinc oxide (ZnO:Al) deposited by sputtering [12, 13], boron-doped zinc oxide (ZnO:B) grown by low-pressure chemical vapor deposition (LPCVD) [14] and fluorine-doped tin oxide ($SnO_2$:F) grown by atmospheric-pressure chemical vapor deposition (APCVD) [15, 16].

Novel materials such as gallium-doped ZnO have been suggested as alternative compounds for use as front TCOs in p-i-n solar cells [17]. Sputtered tin-doped indium oxide $In_2O_3$:$SnO_2$ (ITO) combined with a textured superstrate has also been proposed to fulfill this function [18], and more recently hydrogenated indium oxide $In_2O_3$:H (IOH) [19, 20]. Other approaches have also been introduced to improve the FOM of ZnO:Al front electrodes and, with it, cell performance: for instance, two-step post-deposition annealing treatments, comprising a thermal treatment under a protective layer, result in a decrease in sub-bandgap and free-carrier absorption combined with a remarkable gain in carrier mobility [9, 21, 22].

In wafer-based photovoltaic technologies or for TF-Si solar cells in the substrate (n-i-p) configuration, a metal grid is conventionally combined with the front TCO electrode to aid in lateral conduction [23, 24]; in this case, the front TCO layer usually exhibits relatively high $R_{sq}$ (up to 100 Ω/sq) as its thickness is generally kept thin (around 70 nm) to fulfill the role of antireflective coating [25]. In the present study, we propose to extend this well-established concept to TF-Si solar cells in the superstrate (p-i-n) configuration to relax the constraint on conductance imposed on front TCOs. In particular, this approach was adopted in the framework of the "Nanospec" project, where TCOs with high transparency in the infrared spectral range (and thus moderate conductance) were required to take full advantage of up-converter systems placed behind the cell [26,27].

Here, we show that single-junction microcrystalline silicon ($\mu$c-Si:H) solar cells with excellent electrical properties can be obtained on highly transparent TCO layers of relatively high $R_{sq}$ (up to approximately 30 Ω/sq) fitted with silver (Ag) fingers with optimized density. We demonstrate that this optimum can be well predicted with the optoelectronic device simulator "Advanced Semiconductor Analysis" (ASA). Then, we show that the shadowing losses induced by the Ag fingers can be noticeably reduced by a transparent retroreflective textured foil placed at the air/glass interface. We highlight the relevance of the introduced concept by comparing the cell performance with that of a cell with a state-of-the-art ZnO:Al front electrode.

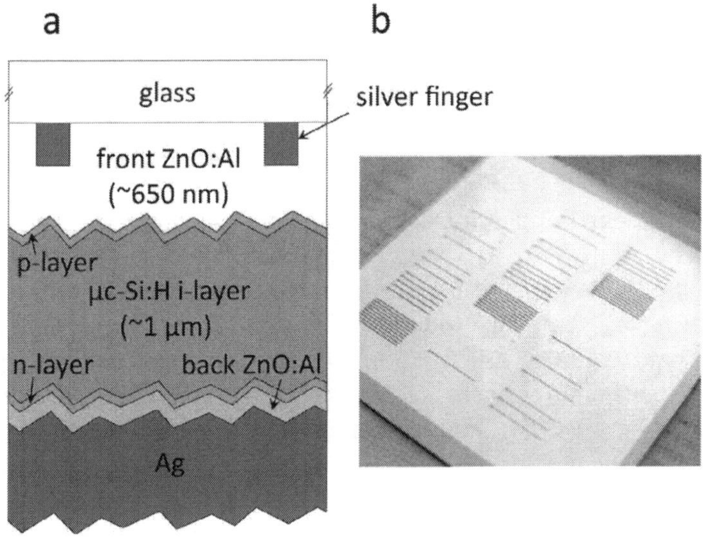


**Figure 1** (a) Schematic cross section of a µc-Si:H p-i-n solar cell deposited on a highly transparent front TCO layer with Ag fingers. (b) Evaporation mask employed to deposit the Ag fingers on the glass superstrate. The line width is varied between 100 µm and 200 µm.

## 2 EXPERIMENTAL DETAILS

The p-i-n $\mu$c-Si:H solar cells were deposited on 1.1-mm-thick glass (Corning Inc.). Figure 1a illustrates the schematic cross section of a cell with Ag fingers. The Ag fingers were evaporated on glass through a stainless steel mask, which was patterned by laser scribing, providing Ag fingers with a width of approximately 100 $\mu$m or 200 $\mu$m and a thickness of around 700 nm. In this study, the height of the fingers was not varied. However, we believe that in practice, for the finger widths considered here ($\sim$100 $\mu$m), the thickness of the fingers could certainly be further reduced, while still providing sufficient conductivity and current-collection ability. The mask was engineered in a way that various finger geometries, i.e. with varying distances between the fingers (see Fig. 1b), could be obtained simultaneously on the same glass substrate. As a front electrode, a ZnO:Al bi-layer with a 1 wt.% aluminum doping concentration was deposited by radio-frequency magnetron sputtering, the first nanometers of ZnO:Al being deposited at room temperature to limit thermal stress in the Ag fingers, and the bulk ZnO:Al layer being deposited at a substrate temperature of around 300 °C. Note that, for cells with and without Ag fingers, the same front TCO was used (by co-deposition). Etching in a 0.5% w/w diluted hydrochloric acid (HCl) solution was carried out to texturize the initially smooth ZnO:Al surface, leading to a final average thickness of approximately 650 nm. The sheet resistance of the resulting ZnO:Al layers is typically around 8 Ω/sq.

In this study, a thermal treatment in vacuum (at 600 °C for 3.5 h to 5.5 h) was performed to increase the transparency of the front ZnO:Al layers fitted with the Ag fingers. A similar treatment at 400 °C did not show any noticeable modification of the opto-electrical properties of ZnO:Al. Annealing at 600 °C is challenging for industrial applications due to the severe cost issues associated to the use of Corning glass and to the thermal treatment itself. In principle, alternative TCO materials to ZnO:Al – with a higher transparency – could have been used in this study. However, one of the advantages of annealing is that it preserves the surface morphology of the front ZnO:Al [28]. It thus enables one to investigate the impact of the ZnO:Al intrinsic optical and electrical properties without considering eventual modifications of the light-trapping properties or of the Si growth conditions associated with a variation of the surface texture.

The Si layers were deposited by plasma-enhanced chemical vapor deposition (PECVD). The boron- and phosphorous-doped (p- and n-) layers and the intrinsic (i-) layer have thicknesses of approximately 20 nm and 1 $\mu$m, respectively. More details on the preparation of our baseline solar cells can be found elsewhere [29]. An 80-nm-thick sputtered ZnO:Al layer and a 700-nm-thick evaporated Ag layer were used together as a back electrode. The cell area (of 1 $cm^2$) is defined by the Ag back reflector, which serves as an etching mask for the underlying ZnO:Al layer.

An antireflective/retroreflective foil from the company DSM was applied to the front side of the device at the air/glass interface. Details on the specifications and working principle of this foil are given later in the text.

The reflectance $R$ and transmittance $T$ of TCOs – from which their absorptance was determined (as $A = 1 - R - T$) – were acquired using a LAMBDA 950 (Perkin Elmer) spectrophotometer equipped with an integrating sphere. Current-voltage ($J$-$V$)measurements were carried out using a dual-lamp solar simulator (Wacom) under standard test conditions (AM1.5G spectrum, 1000 W/$m^2$, 25 °C).

## 2.1 Simulations with ASA and model calibration

To support the measurements, full one-dimensional numerical device simulations were performed. For this purpose, we used the commercial optoelectronic device simulation software ASA [30]. The simulations account for two important aspects, namely the shadowing by the fingers and the resistive losses in the front electrode. The optical response of the solar cells was calibrated with absorptance measurements of a flat bifacial $\mu$c-Si:H p-i-n device, similar to the devices used here but without the back reflector [26]. Owing to the low coverage of the fingers (well below 5%), the additional light scattering presumably resulting from the surface modulation around the fingers has likely only a minor impact on the overall optical performance (and thus on $J_{sc}$). Also, due to the low height-to-width aspect ratio (<0.01) of the fingers, additional shadowing caused by illumination under oblique incidence is most probably negligible. Therefore, we assume that the 1-D model used here to account for the optical impact of the fingers already provides good estimation. The electrical parameters were derived from *J-V* characteristics of the device in the dark and under AM1.5G illumination. The illuminated *J-V* characteristics are therefore a superposition of *J-V* characteristics from areas in the dark (those under the fingers, $J_d$) and areas that receive illumination (everywhere else, $J_i$). The current is hence calculated as:

$$J = \frac{d_{\mathrm{Ag}}}{d} J_d + \frac{d_{\mathrm{Ag-Ag}}}{d} J_i,$$

where $d_{\mathrm{Ag}}$ is the width of a finger, $d_{\mathrm{Ag-Ag}}$ the gap between two adjacent fingers, and $1/d$ the finger density. Note that $d = d_{\mathrm{Ag-Ag}} + d_{\mathrm{Ag}}$.

Solving the differential equation for the electrostatic potential between two fingers in one dimension yields the contact series resistance

$$R_s = R_{s0} + \frac{R_{\mathrm{sq}}^{\mathrm{ZnO}} d_{\mathrm{Ag-Ag}}^2}{8},$$

where $R_{s0}$ is a constant offset (used here as a fitting parameter and set to a value of 0.08 Ω), and $R_{\mathrm{sq}}^{\mathrm{ZnO}}$ the sheet resistance of the ZnO:Al front contact [26]. The variation of $1/d$ then yields – through $R_s$ – the corresponding change in fill factor (*FF*), short-circuit current density ($J_{sc}$) and efficiency of the solar cell.

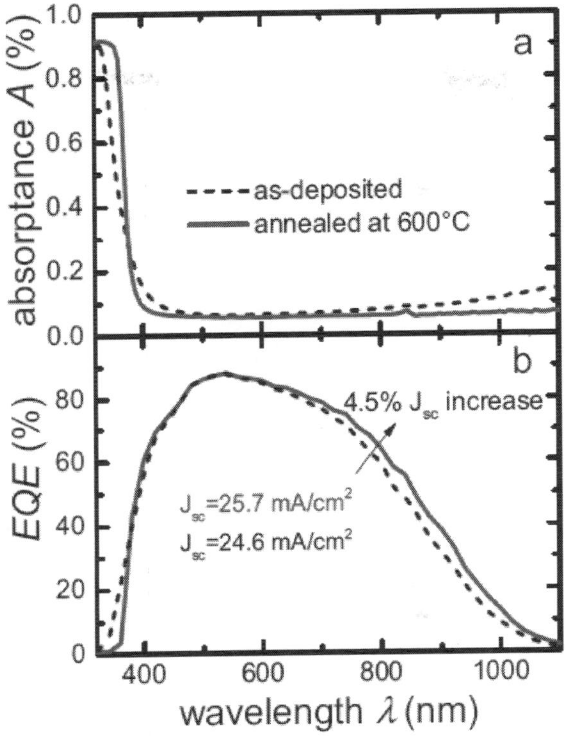


**Figure 2**.(a) Absorptance *A* as a function of wavelength of glass/ZnO:Al superstrates with and without a post-deposition annealing treatment at 600 °C for 5.5 h. (b) External quantum efficiency *EQE* of the solar cells deposited on these superstrates. The corresponding photocurrent values ($J_{sc}$) are given in the graph

## 3 RESULTS AND DISCUSSION

Figure 2 illustrates the impact of an annealing treatment on the optical properties of ZnO:Al and on the spectral performance of a complete solar cell. Below 380 nm, the ZnO:Al layer shows a higher absorptance (lower transparency) after annealing at 600 °C for 5.5 h (Fig. 2a). This effect is explained by the reduced band filling – and associated narrowing of the optical bandgap (Burstein-Moss effect) – caused by the decrease in free-carrier density upon thermal treatment [31, 32]. Above 700 nm, the reduction of the free-carrier density leads to a noticeable increase in layer transparency, which is mostly dictated by the shift of the free-electron plasma resonance towards longer wavelengths [33, 34]. Finally, sub-bandgap absorption in the spectral range below the fundamental absorption edge, i.e. from 380 nm to 500 nm, also decreases, as already reported in reference [9]. In accordance with the absorptance data of the glass/ZnO:Al superstrates, the external quantum efficiency (*EQE*) of the cell deposited on the annealed superstrate surpasses that of the reference cell in the near-infrared (NIR) wavelength region; this *EQE* gain compensates for the slight optical loss below 380 nm, such that the photocurrent increases from 24.6 mA/cm$^2$ to 25.7 mA/cm$^2$ (by 4.5%). The lower absorptance of the treated ZnO:Al between 380 nm and 500 nm does not translate into an *EQE* gain; the optical behaviour of single layers on glass strongly differs from that of more complex optical systems like complete solar cells and further investigation would be needed to understand this effect.

Annealing ZnO:Al in vacuum at 600 °C typically results in an increase in sheet resistance, due to the decreased carrier concentration and to a lower Hall mobility [10, 33]. In the literature, the latter effect is attributed to residual oxygen gas in the vacuum chamber, leading to oxygen adsorption – and thus probably to an increased barrier height – at the grain boundaries of ZnO:Al [35, 36].

After 3 h of annealing at 600 °C in vacuum, the ZnO:Al layer becomes highly transparent, even though $R_{sq}$ remains below 25Ω/sq. Figure 3 shows the *FF*, $J_{sc}$ and efficiency of 1-cm$^2$ full-area $\mu$c-Si:H cells deposited on thermally treated front ZnO:Al layers (with a $R_{sq}$ of around 20 Ω/sq) equipped with 200-$\mu$m-wide Ag fingers. The finger density is varied from 0 (i.e. with no finger) to 10 fingers per cm. The $V_{oc}$ exhibits almost no variation upon modification of the finger density (not shown here). Owing to the relatively thick ZnO:Al layer (of approximately 650 nm), the abruptness of the transition step at the finger edges might be strongly attenuated, preventing the occurrence of too severe structural defects in the subsequent layers that would deteriorate cell performance. The *FF* first rapidly increases from 61% to 72% by adding a single finger (see Fig. 3a, solid black line) and then saturates at a value of 74% for higher finger densities. Note that the strong benefit observed when adding a single finger is explained by the rather long distance (of around 3 mm) separating the cell edges and the bus bar where the contact is made. As expected, the $J_{sc}$ decreases linearly with increasing Ag finger density (Fig. 3b) due to increased shadowing. The highest $FF \times J_{sc}$ product, and consequently the highest efficiency, is obtained with two fingers per cm (Fig. 3c, solid black line). The existence of this optimal configuration is in good agreement with theoretical predictions obtained with ASA (Fig. 3, dashed lines).

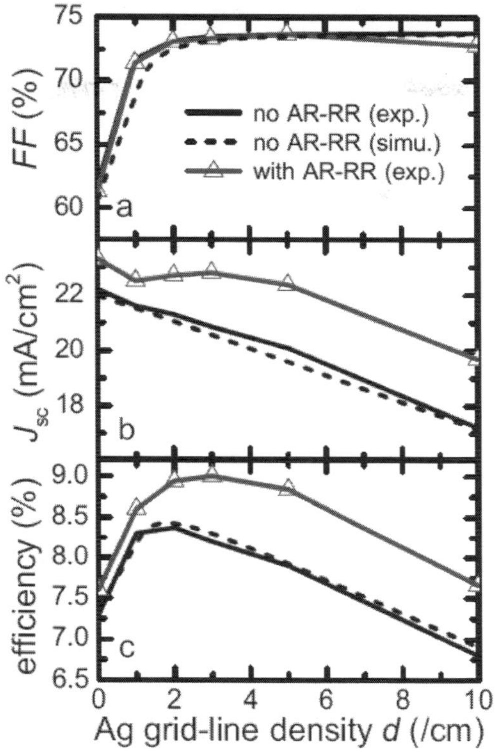


**Figure 3**. Experimental and calculated parameters of $\mu$c-Si:H solar cells deposited on thermally treated front ZnO:Al layers fitted with Ag fingers of varying density. The curves with triangles represent parameters of cells with an antireflective/retroreflective (AR-RR) foil. Note that finger densities *d* of0 $<d<$ 1 make sense only for the simulated data since the experimental parameters extracted here were measured on cells with an active area of 1 × 1 cm$^2$. The case $d = 0$ represents the cell with no Ag fingers.

To improve light coupling into solar cells, an antireflective coating is traditionally applied at the air/glass interface. Geometric textures with feature sizes well above the wavelength of the incoming light can be introduced at this interface to fulfill this function [37, 38]. The antireflective effect is based on the rebound of the incident light at the facets of the texture, as illustrated in Figure 4 (effect 1) for pyramidal features. One particularity of such textures is that they also permit some of the light reflected at the first interfaces (glass/ZnO:Al, ZnO:Al/Si) or at the Ag fingers to be redirected towards the absorber layer (see Fig. 4, effect 2). In the present study, the antireflective/retroreflective (AR-RR) texture is of particular importance since it mitigates the optical (shadowing) losses induced by the fingers by effectively retroreflecting the light back into the solar cell. Here, the textured cover consists of an adhesive flexible polymeric foil, from the company DSM, with a refractive index similar to that of glass, attached directly to the glass superstrate. The positive impact of the AR-RR foil on the device performance is demonstrated in Figure 3 (lines with triangles): the AR-RR foil leads to a substantial gain in $J_{sc}$ (Fig. 3b). Note that the relative gain in $J_{sc}$ obtained with the foil increases with finger density, which underlines the effectiveness of its retroreflective ability. As a result of this effect, a new optimum of the finger geometry is found with the AR-RR foil, which gives a maximum power conversion efficiency of approximately 9% with three fingers (per cm).

**Table 1** *J-V* parameters of cells (average of the three best), with and without an AR-RR foil, deposited on a reference ZnO:Al layer (ref.) and on a ZnO:Al layer annealed at 600 °C for 3.5 h (ann.) fitted with Ag fingers.

| | Efficiency (%) | $FF$ (%) | $J_{sc}$ (mA/cm$^2$) | $V_{oc}$ (mV) |
|---|---|---|---|---|
| ref. | 8.5 | 70.3 | 23.1 | 525 |
| ref. + AR-RR | 8.7 | 70.1 | 23.7 | 525 |
| ann. + fingers | 8.7 | 71.9 | 22.9 | 530 |
| ann. + fingers + AR-RR | 9.1 | 71.9 | 23.8 | 530 |

By reducing the width of the fingers to 100 $\mu$m (i.e. by 50%) and by choosing a density of three fingers per cm, we were able to further increase the cell performance. Figure 5 depicts the *J-V* characteristics of a cell deposited on such a front electrode and of a reference cell, co-deposited on a bare, untreated ZnO:Al layer. The corresponding cell parameters with and without the AR-RR foil are listed in Table 1.

The major difference resides in the *FF*: The annealed ZnO:Al layer fitted with Ag fingers provides a higher *FF*, by around 2.5% (rel.), than the reference ZnO:Al electrode. This gain underlines the beneficial role of the metal fingers, which more than compensate for the increased Ohmic losses in the annealed ZnO:Al layer. For cells without an AR-RR foil, a slightly smaller $J_{sc}$ is measured for the front electrode with Ag fingers, due to shadowing. The application of the AR-RR foil leads to a $J_{sc}$ increase of around 2.6% for the reference cell and 4% for the cell with fingers. Notably, a slightly higher $V_{oc}$ is measured for the cell with the annealed ZnO:Al layer (accounting for a relative efficiency gain of less than 1%). Overall, cells with the AR-RR foil are significantly improved by substituting the reference ZnO:Al layer with a highly transparent front electrode fitted with Ag fingers: The efficiency increases from 8.7% to 9.1%.

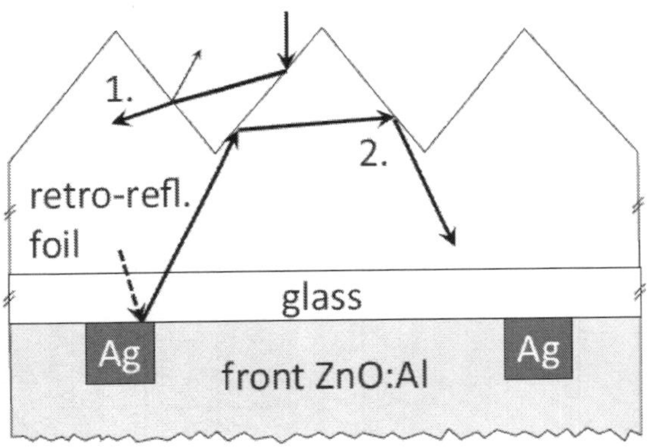


**Figure 4.**Illustration of the geometric-optics interaction between a ray of light and the anti-reflective/retroreflective foil. Effect 1: the fraction of incoming light that is not transmitted into the foil upon the first incidence will hit a second facet, resulting in a higher fraction of transmitted light. Effect 2: light reflected out of the cell, in particular at the Ag fingers, is redirected back towards the active layers of the cell via a double rebound of the light beam at the textured-foil/air interface

To conclude, we demonstrated that the combination of a highly transparent front TCO with metal fingers is a promising approach to obtain a better compromise between the optical and electrical constraints of TF-Si solar cell front electrodes. We highlighted the utility of this concept for 1-$\mu$m-thick $\mu$c-Si:H solar cells deposited on thermally treated ZnO:Al layers with Ag fingers, where a noticeable gain in *FF* has been measured as compared to reference cells deposited on bare, untreated ZnO:Al. By introducing a millimeter-scale textured foil providing an antireflective and retroreflective effect at the air/glass interface, we have been able to offset the impact of the shadowing losses induced by the fingers. By extending this approach to other TCO materials such as sputtered ZnO:Al with a lower doping concentration or to IOH or LPCVD-ZnO and by using optimized geometries for the metal fingers, we anticipate further advances in the efficiency of TF-Si solar cells in the single- or multi-junction device configuration.

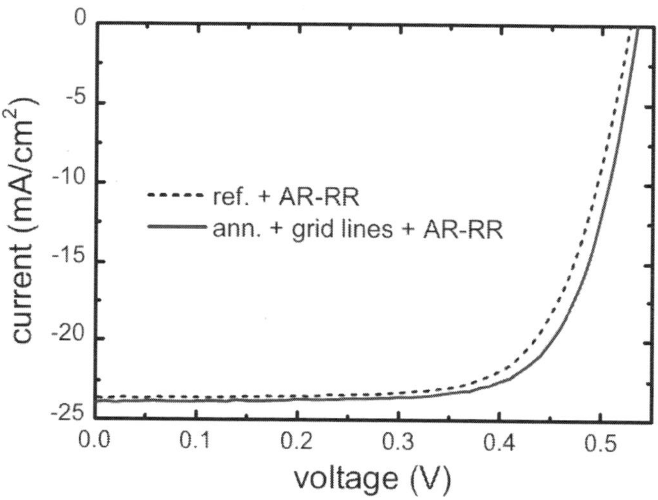


**Figure 5**. *J-V* curves of 1-$\mu$m-thick $\mu$c-Si:H solar cells deposited on a reference ZnO:Al layer and on a ZnO:Al layer annealed at 600 °C for 3.5 h equipped with Ag fingers. The width of the fingers is approximately 100 $\mu$m and the density is 3 fingers per cm. Both cells are covered with an AR-RR foil.

## ACKNOWLEDGMENTS

We gratefully acknowledge Jürgen Hüpkes, Stephan Haas, Carolin Ulbrich, and Juraj Hotovy for constructive discussions. This work was carried out in the framework of the project "Nanospec", funded by the European Community's Seventh Framework Programme (FP7/2007-2013) under grant agreement no 246200.

## REFERENCES


1. E. Yablonovitch, G.D. Cody, IEEE Trans. Electron Devices **29**, 300 (1982)
2. H.W. Deckman et al., Appl. Phys. Lett. **42**, 968 (1983)
3. O. Kluth et al., Thin Solid Films **351**, 247 (1999)
4. M. Ermes et al., J. Appl. Phys. **113**, 073104 (2013)
5. J.I. Owen et al., Phys. Stat. Sol. A **208**, 109 (2011)
6. J. Bailat et al., in *Proceedings of the 25th European Photovoltaic Solar Energy Conf./5th World Conf. Photovoltaic Energy Convers, Valencia, 2010*, p. 2720
7. M. Python et al., Prog. Photovolt.: Res. Appl. **18**, 491 (2010)
8. S. Hänni et al., IEEE J. Photovolt. **99**, 1 (2012)
9. F. Ruske et al., J. Appl. Phys. **107**, 013708 (2010)
10. T. Minami et al., J. Vacuum Sci. Technol. A **17**, 1822 (1999)
11. R.G. Gordon, MRS Bulletin **25**, 52 (2000)
12. A. Lambertz et al., Sol. Energy Mater. Sol. Cells **119**, 134 (2013)
13. S. Kim et al., Sol. Energy Mater. Sol. Cells **119**, 26 (2013)
14. M. Boccard et al., IEEE J. Photovolt. **2**, 229 (2012)
15. M. Kambe et al., in *Proceedings of the 3rd World Conf. Photovoltaic Energy Conversion, Osaka, 2003*, Vol. 2, p. 1812
16. J. Krc et al., Thin Solid Films **518**, 3054 (2010)
17. K.C. Lai et al., Sol. Energy Mater. Sol. Cells **94**, 397 (2010)
18. C. Battaglia et al., Appl. Phys. Lett. **96**, 213504 (2010)
19. C. Battaglia et al., J. Appl. Phys. **109**, 114501 (2011)

20. T. Koida et al., Jpn J. Appl. Phys. **28**, L685 (2007)

21. M. Wimmer et al., Thin Solid Films **520**, 4203 (2012)

22. S. Neubert et al., Prog. Photovolt.: Res. Appl. **22**, 1285 (2014)

23. J. Geissbühler et al., IEEE J. Photovolt. **4**, 1055 (2014)

24. H. Sai et al., in *Proceeding of the MRS conference, San Francisco, 2013*, Vol. 1536, p. 3

25. E. Kobayashi et al., in *Proceedings of the 28th European Photovoltaic Solar Energy Conference and Exhibition, Paris, 2013*, p. 691

26. T.C.M. Müller et al., Energy Procedia **10**, 76 (2011) J.C. Goldschmidt et al., in *Renewable Energy and the environment* (Optical Society of America, 2013), paper PT3C.2

27. J. Owen, Ph.D. thesis, RWTH Aachen University, 2011

28. B. Rech et al., Thin Solid Films **511**, 548 (2006)

29. B.E. Pieters et al., J. Appl. Phys. **105**, 044502 (2009)

30. B.E. Sernelius et al., Phys. Rev. B **37**, 10244 (1988)

31. E. Burstein, Phys. Rev. **93**, 632 (1954)

32. M. Berginski et al., Thin Solid Films **516**, 5836 (2008)

33. A. Pflug et al., Thin Solid Films **455**, 201 (2004)

34. S. Takata et al., Thin Solid Films **135**, 183 (1986)

35. T. Minami, H. Sato, K. Ohashi, T. Tomofuji, S. Takata, J. Cryst. Growth **117**, 370 (1992)

36. C. Ulbrich, Prog. Photovolt.: Res. Appl. **21**, 1672 (2013)

37. J. Escarre et al., Sol. Energy Mat. Sol. Cells **98**, 185 (2012)

# Chapter 10

## Recrystallized thin-film silicon solar cell on graphite substrate with laser single side contact and hydrogen passivation

**Da Li[1a], Stephan Wittmann[1], Thomas Kunz[1], Taimoor Ahmad[1], Nidia Gawehns[1], Maik T. Hessmann[1], Jan Ebser[2], Barbara Terheiden[2], Richard Auer[1] and Christoph J. Brabec[1,3]**

[1] Bavarian Center for Applied Energy Research (ZAE Bayern), Haberstr. 2a, 91058 Erlangen, Germany

[2] Department of Physics, University of Konstanz, Box 676, 78457 Konstanz, Germany

[3] Institute of Materials for Electronics and Energy Technology (i-MEET), University of Erlangen-Nuremberg, Martensstr. 7, 91058 Erlangen, Germany

## ABSTRACT

Laser single side contact formation (LSSC) and the hydrogen passivation process are studied and developed for crystalline silicon thin film (CSiTF) solar cells on graphite substrates. The results demonstrate that these two methods can improve cell performance by increasing the open circuit voltage and fill factor. In comparison with our previous work, we have achieved an increase of 3.4% absolute cell efficiency for a 40 $\mu$m thick 4 $cm^2$ aperture area silicon thin film solar cell on graphite substrate. Current density-voltage (*J*-*V*) measurement, quantum efficiency (QE) and light beam induced current (LBiC) are used as characterization methods.

For crystalline silicon thin film (CSiTF) solar cells on the foreign substrates, a recrystallization process plays an important role in enlarging the size of the silicon grains in order to reduce the density of the electrically active defects and increase the cell efficiencies, and hence the electron-beam recrystallization [1, 2], zone melting recrystallization (ZMR) [3, 4] and laser recrystallization [5, 6, 7, 8, 9] are widely developed [10, 11, 12, 13]. Foreign substrate materials, such as ceramics [14, 15, 16] and graphite [17], are generally used for CSiTF cell fabrication with the high temperature approach. Schillinger et al. [18] described a CSiTF cell concept on zircon ceramic substrates using ZMR reached 8.1% conversion efficiency. Graphite substrates can be processed at high temperature. Moreover, they are available at high purity and with thermal expansion characteristics similar to silicon. While standard graphite types will not meet the cost requirements, developments for low-cost types such as biogenic substrates are in progress [19].

In previous work, we have reported a best laboratory cell efficiency of 6.8% with the 4 $cm^2$ aperture area on the graphite substrate [20]. That cell concept is shown in Figure 1a. The laser edge isolation (LEI) technique was applied to avoid the parasitic electrical connection between the front and back contacts. This technique was proved to be a convenient and accurate method instead of the plasma etching process. By using the LEI technique, the best parallel resistance, approximately 1.7 $\Omega m^2$, was achieved.

Much effort has been devoted to the metallization of CSiTF solar cells on foreign substrates. Stocks et al. [21] presented the concept of CSiTF solar cells with base front-contacts. Hebling et al. [22] realized the concept by using photolithography. Meanwhile, Lüdemann et al. [23] realized base front-contacts silicon solar cells on a SiC intermediate layer with a graphite substrate. Furthermore, Rachow et al. [24] named the concept as single side contact formation. Different microstructures of single side concepts can be formed using photolithography [21] or reactive ion etching (RIE) [25]. However, the photolithography is an expensive technology, whereas a typical RIE system consists of a vacuum chamber and a plasma generation system. Due to the development of laser techniques, laser processes are nowadays widely used in thin film solar cell fabric processing. Such techniques can now be used to complete a solar cell with single side contact formation in fewer steps.

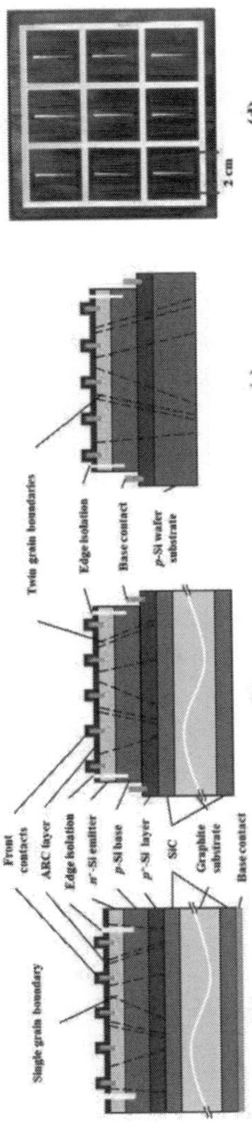


**Figure 1.** The three different Si thin-film solar cell designs were studied. (a) The cell in previous work is based on front and back contacts. (b) The new design in this paper is based on laser single side front contacts (LSSC). (c) The reference cells are based on p-type Si wafer. The layers are not to scale. (d) Top view of batch Gra. C, on which are the nine cells with the graphite substrate. The cell No. 5 is in the center.

In this paper, we developed the cell concept on graphite substrates as shown in Figure 1b. The current density-voltage (*J-V*) curves and quantum efficiency (QE) measurements show the improvements of open circuit voltage ($V_{oc}$), fill factor (*FF*), cell efficiency ($\eta$), and quantum efficiency due to the hydrogen bulk passivation process [26, 27] and laser single side contact (LSSC) formation. The combined solar cell process is carried out at a low temperature (<350 °C). The diffusion length was characterized using light beam induced current (LBiC) measurement and simulated by PC1D [28]. Here we report our 10.2% cell efficiency of a 40 $\mu$m thick and 4 $cm^2$ aperture area CSiTF solar cell on graphite substrate.

# 1 EXPERIMENTAL

## 1.1 Solar cell design and processing

Three different cell designs (Figs. 1a–1c) were studied. The designs illustrated in Figures 1a and 1b are based on the use of graphite as a foreign substrate, with different concepts of base contact formation. In designing Figure 1c, a multi-crystalline Si-wafer was chosen as a base material for reference purposes, because recrystallized silicon layers on graphite substrates have a similar grain size to typical multi-crystalline Si-wafers. Results from the design of Figure 1a have been described in a previous work [20], while the other two designs are investigated in this work.

For the foreign substrates (Figs. 1a and 1b), high purity graphite substrates ("FP479", Schunk Kohlenstofftechnik GmbH) with a size of 10cm × 10cm × 2mm were used. The SiC layer with a thickness of 10 $\mu$m was deposited using a hot-wall chemical vapor deposition, which covered both surfaces of the substrate to prevent the diffusion of impurities. Then, a $p^+$-Si layer (acceptor concentration of 4 × $10^{18}$$cm^{-3}$) approximately 20 $\mu$m thick was deposited on top of the SiC layer using a convection-assisted chemical vapor deposition (CoCVD) [29]. The $p^+$-Si layers served as a seed and back surface field layer. The ZMR process was applied to enlarge the size of the silicon grains from the micrometer to the millimeter range. Thus, a random texture surface was formed due to the various crystalline silicon directions [4].

For the reference substrate (Fig. 1c), the $p^+$-Si layer had the concentration of 4 × $10^{18}$$cm^{-3}$ and was epitaxially grown directly onto the wafer substrate.

Further processing was the same for both the foreign substrate and the Si-wafer substrate. A 20 $\mu$m thick epitaxial *p*-Si base layer was applied over the $p^+$-Si layer. This *p*-Si layer had a boron doping concentration of about 2 × $10^{16}$cm$^{-3}$. The $n^+$-Si emitter was formed by spin on doping (SOD) of a phosphorous solution followed by rapid thermal processing (RTP) in a furnace [30]. The sheet resistance is in the range from 80 to 120 Ω/square. The laser edge isolation process was applied directly after the removal of the phosphorous glass [20]. The metallization of the front contacts were formed by electron beam evaporation of Ti, Pd and Ag (30 nm, 30 nm, and 5 $\mu$m thick, respectively). Finally, using plasma-enhanced chemical vapor deposition (PECVD) we deposited a silicon nitride layer with a thickness of 75 nm, which served as an antireflection coating. The total thickness of the silicon layers was about 40 $\mu$m. As shown in Figure 1d, nine cells, each with a size of 2 cm × 2cm, have been fabricated on the 10 cm × 10cm graphite substrate.

**Table 1** . Sample list. $H^+$ denotes the hydrogen passivation process.

| Batches | Substrate | Contacts | $H^+$ [Yes/No] | Duration (min) |
|---|---|---|---|---|
| Gra. A | Graphite | Both sides | No | – |
| Gra. B | Graphite | LSSC | No | – |
| Gra. C | Graphite | LSSC | Yes | 50 |
| Waf. A | *p*-Si wafer | Both sides | No | – |
| Waf. B | *p*-Si wafer | LSSC | No | – |
| Waf. C | *p*-Si wafer | LSSC | Yes | 50 |
| Waf. D | *p*-Si wafer | LSSC | Yes | 50 |
| Waf. E | *p*-Si wafer | LSSC | Yes | 50 |

## 1.2 Sample batches

As shown in Table 1, we prepared and studied eight different batches for the comparison of LSSC formation and hydrogen passivation process. The batches based on the graphite substrates are named Gra. A, B, and C, whereas the reference batches based on the *p*-type Si-wafers are named Waf. A, B, C, D, and E. The batches denoted by different capital letters correspond to the different processing sequences as shown in Figures 2 and 3:

- Gra. A and Waf. A are the samples with base contacts on the backof the solar cells. They are based on the design of Figure 1a. Thebase contact is formed by a 5 $\mu$m thick aluminum layer, which was deposited on the back of the substrate. Hydrogen passivation is not used in these batches.

- Gra. B and Waf. B's base contacts were formed from the front using LSSC. They are based on the design of Figure 1b, but without the hydrogen passivation process.

- Gra. C and Waf. C–E are with LSSC and the hydrogen passivation process. Gra. C is based on the design of Figure 1b. Waf. C–E are based on the design of Figure 1c.

## 1.3 Laser single side contact

Based on the development of laser technology, we fabricated the base contacts on the front using a laser process, i.e. laser single side contacts instead of the Al back base contacts. Each cell from batches Gra. A and Waf. A had an aluminum base contact, which was evaporated on the back. Here we used a Nd:$YVO_4$ laser (Rofin Power Line LP20, wavelength 1064 nm) to make approximately 30 $\mu$m deep trenches into the samples down to the $p^+$-Si layer and to obtain 3 mm wide contact stripes around the cells. After the laser trenching and LEI process, we evaporated the emitter contacts and the base contacts on the front at the same time and with the same metals Ti, Pd, and Ag.

## 1.4 Hydrogen passivation

The bulk recombination centers in the CSiTF solar cells can be restrained using a hydrogen passivation process, which is dependent on time and temperature. According to reference [27], the hydrogen passivation process needs less process time at high temperatures than at low temperatures. If the process temperature is 600 °C, the necessary reaction time is only about 5 min. In our case, the samples were put into the chamber at 350 °C for 50 min (microwave power: 1000 W, pressure: 0.04 mbar and hydrogen as the precursor). Experiments were carried out using a PECVD system of Roth and Rau AK1000 with microwave excitation frequency at 2.45 GHz.

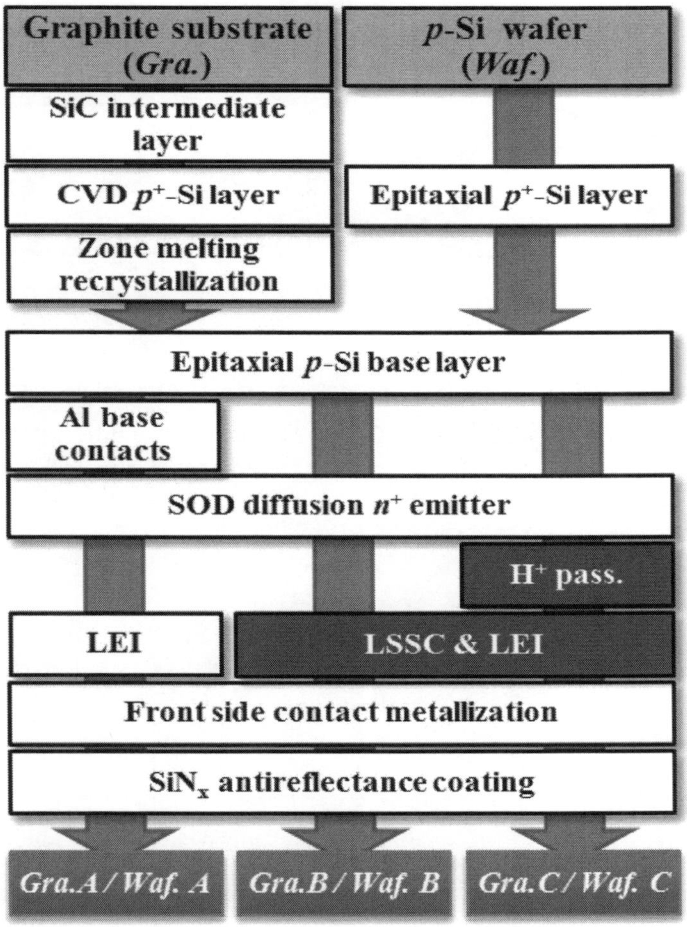


**Figure 2**. Processing sequence of the cells. Eight solar cell batches are prepared. Gra. denotes the cells on graphite substrate and Waf. denotes the cells on wafers as reference. Laser edge isolation (LEI), laser single side contact (LSSC) and hydrogen bulk passivation (H+ pass.) were applied in the cell process. Batches Gra. A–C on graphite substrates, reference batches Waf. A–C on Si wafers with the same processing after ZMR as Gra. A–C.

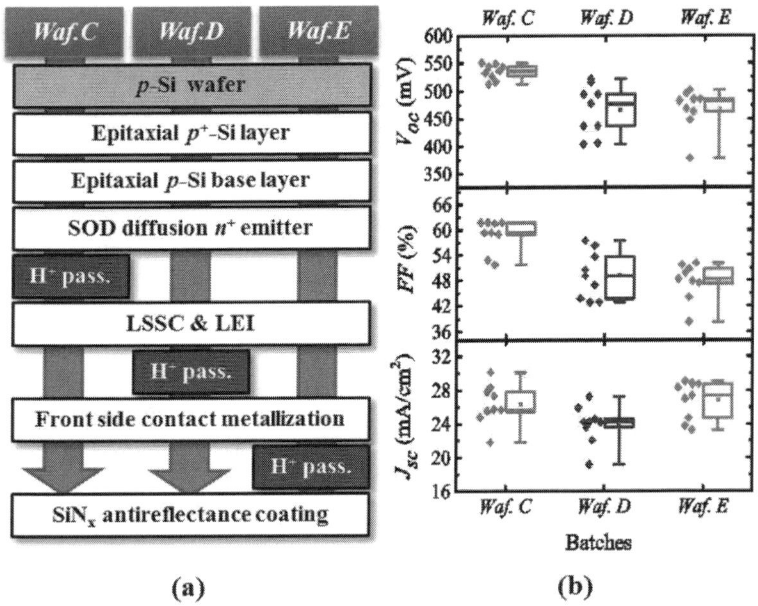


**Figure 3.** Reference batches Waf. C–E on Si wafer with a different sequence of hydrogen passivation. (a) Processing sequence of the cells. (b) Open circuit voltage Voc (top), fill factor FF (middle) and current density Jsc (bottom) due to the three different hydrogen passivation processes.

**Table 2** . Cell parameters. Open circuit parallel resistance ($R_p$), series resistance ($R_s$), voltage ($V_{oc}$), current density ($J_{sc}$), fill factor ($FF$), and efficiency ($\eta$).

| Batches | Avg. $R_p$ ($\Omega$ m$^2$) | Avg. $R_s$ ($\Omega$ cm$^2$) | Avg. $V_{oc}$ (mV) | Avg. $J_{sc}$ (mA/cm$^2$) | Avg. $FF$ (%) | Avg. $\eta$ (%) | Best $\eta$ (%) |
|---|---|---|---|---|---|---|---|
| Gra. A | 1.53 | 1.52 | 426 | 21.3 | 56.7 | 5.18 | 6.79 |
| Gra. B | 1.97 | 1.06 | 469 | 234 | 66.5 | 7.33 | 7.81 |
| Gra. C | 2.47 | 1.05 | 499 | 262 | 67.7 | 8.86 | 10.2 |
| Waf. A | 1.08 | 2.22 | 450 | 255 | 49.0 | 5.74 | 7.68 |
| Waf. B | 2.51 | 2.77 | 501 | 250 | 57.0 | 7.18 | 8.19 |
| Waf. C | 3.18 | 2.85 | 534 | 263 | 58.7 | 8.29 | 10.1 |
| Waf. D | 0.33 | 3.24 | 465 | 23.9 | 49.2 | 5.52 | 7.77 |
| Waf. E | 0.03 | 2.21 | 467 | 26.7 | 47.7 | 5.97 | 7.14 |

The process sequence is critical for hydrogen passivation. As shown in Figure 3, in order to determine the best fabrication sequence of the hydrogen passivation, we prepared and compared three different process sequences. For this study, the reference system on wafer substrates was used in order to exclude influences from varying defect structures, as they may result from the recrystallization process. The hydrogen passivation process was included after

- the SOD diffusion processing (Waf. C);
- the laser trenching and LEI (Waf. D);
- the metallization processing (Waf. E).

## 1.5 Cell characterization

Internal quantum efficiency (IQE) and reflectance are measured using the pv-tools solar cell analysis system LOANA at the University of Konstanz and using the Enlitech solar cell analysis system QE-R at i-MEET. Light beam induced current (LBiC) measurements were carried out by using Semilab WT2000, which has four light sources (976 nm, 951 nm, 846 nm and 662 nm). The diffusion length was calculated from the LBiC measurements at various wavelengths.

# 2 RESULTS AND DISCUSSION

## 2.1 J-V characterization

Table 2 lists the parameters obtained from the illuminated and dark *J-V* measurements. We arranged the $J_{sc}$, $V_{oc}$ and *FF* data of all the samples in the form of a box chart, as shown in Figure 3b. These results were used to identify the best hydrogen passivation sequence, which is described in Section 2.1.1. Furthermore, we observed significant improvement in $V_{oc}$ and *FF*with hydrogen bulk passivation and LSSC formation in Figure 4, which is described in Section 2.1.2.

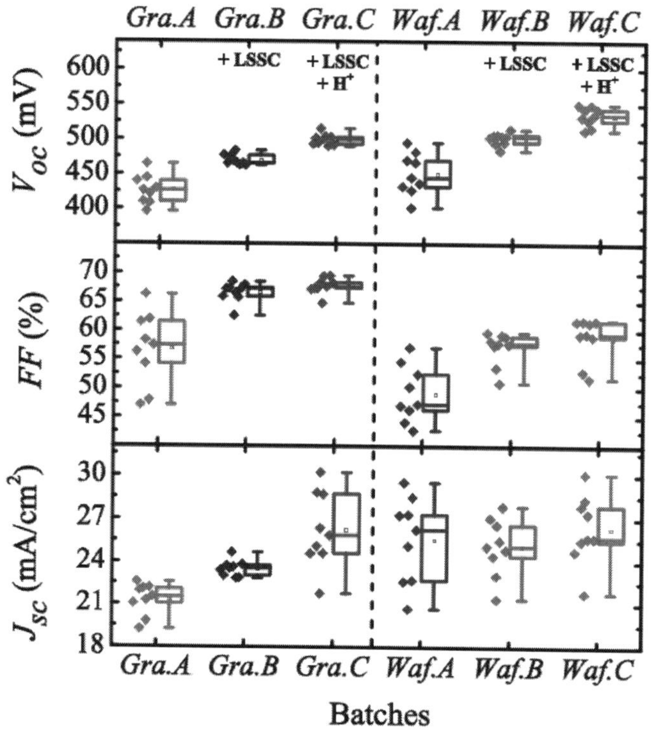


**Figure 4.** Voc (top), FF (middle) and Jsc (bottom) due to the both-side contact, LSSC formation (i.e. +LSSC), and hydrogen passivation processes (i.e. +H+).

## 2.1.1 Hydrogen passivation sequence on reference batches

Comparing the results of these three batches (Waf. C–E) in Figure 3b and Table 2, we found the best hydrogen passivation sequence is the hydrogen bulk passivation process of Waf. C. This batch has the highest average values of $R_p$, $V_{oc}$, and *FF*among the three batches.

In the case of Waf. D, the hydrogen passivation was applied after the laser trenching and LEI process. This process sequence resulted in lower cell performance than Waf. C, which is most probably due to laser induced defects. Slaoui et al. [31] reported the hydrogen passivation of laser induced defects, which demonstrated that the quality of the hydrogen passivation highly depends on the temperature of the passivation process.

In the case of Waf. E, the contacts were already evaporated before the hydrogen passivation process. As shown in Table 2, this process sequence resulted in extremely lower average parallel resistance ($\overline{R_p}$) than Waf. C, which further resulted in a low average ($\overline{FF}$). This was assumed due to the fact that the front contacts (Ag/Pd/Ti) penetrated through the thin emitter layer into the base layer during the passivation process at 350 °C [32].

Accordingly, the best hydrogen passivation sequence was made directly after the phosphor diffusion process, which was also applied to fabricate Gra. C.

### 2.1.2 Batches with hydrogen passivation and LSSC on graphite substrates

In order to confirm the improvement by LSSC formation and hydrogen passivation, we further compared the results of the graphite substrate batches (Gra. A–C) and the reference batches (Waf. A–C). There are three different formations as shown in Figure 2.

As shown in Figure 4, both the graphite-based (Gra.) and reference (Waf.) batches exhibit similar variations in $\overline{V_{oc}}$ and $\overline{FF}$, which can be described by:

$$\overline{V}_{oc}\,(\text{Gra. A}) < \overline{V}_{oc}\,(\text{Gra. B}) < \overline{V}_{oc}\,(\text{Gra. C}) \qquad (1)$$

$$\overline{V}_{oc}\,(\text{Waf. A}) < \overline{V}_{oc}\,(\text{Waf. B}) < \overline{V}_{oc}\,(\text{Waf. C}) \qquad (2)$$

$$\overline{FF}\,(\text{Gra. A}) < \overline{FF}\,(\text{Gra. B}) < \overline{FF}\,(\text{Gra. C}) \qquad (3)$$

$$\overline{FF}\,(\text{Waf. A}) < \overline{FF}\,(\text{Waf. B}) < \overline{FF}\,(\text{Waf. C}). \qquad (4)$$

Those inequalities indicate that the batches with LSSC formation have better performance than the batches with both sides contact formation. The possible reason is that the LSSC formation had less recombination losses than the both-side contact formation, since the collected current bypassed the intermediate layer and the graphite substrate. Moreover, the hydrogen passivation reduced the bulk recombination, and therefore the batches, with both LSSC formation and hydrogen passivation applied have the best performance in $\overline{V_{oc}}$ and $\overline{FF}$. We also observed that the $\overline{J_{sc}}$ of the cells based on graphite substrates had the same trend:

$$\overline{J}_{sc}\,(\text{Gra. A}) < \overline{J}_{sc}\,(\text{Gra. B}) < \overline{J}_{sc}\,(\text{Gra. C}). \qquad (5)$$

We observed that the values of $\overline{J_{sc}}$ were improved using the LSSC formation by comparison $\overline{J_{sc}}$ (Gra. A) and $\overline{J_{sc}}$ (Gra. B). We also observed $\overline{J_{sc}}$ (Gra. C) was the highest due to the applied LSSC and hydrogen bulk passivation. However, the values of $\overline{J_{sc}}$ of the reference cells based on mc-Si wafers did not have the same trend. According to the process sequence, as shown in Figure 2, the reason can be that the reference cells were with neither an intermediate layer nor a ZMR process. Moreover, the back surfaces of the reference cells were not passivated. Therefore, the improvement using hydrogen bulk passivation cannot counteract the detrimental effect of the back surface recombination.

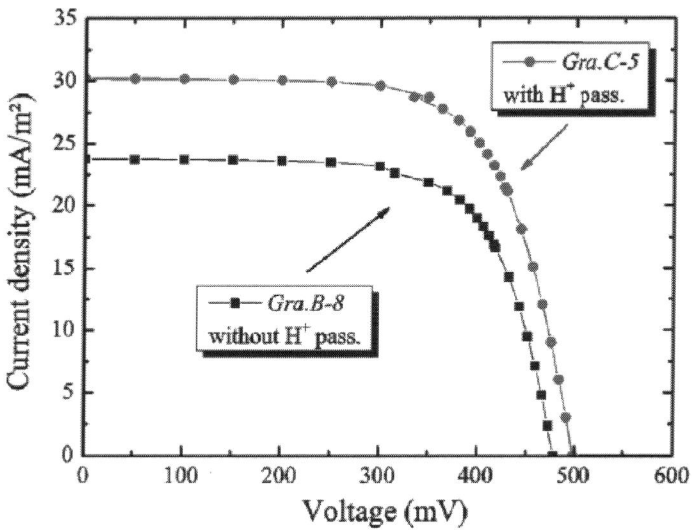


**Figure 5**. *J-V* curves of the cells Gra. B-8 and Gra. C-5. (i.e. LSSC formation without (Gra. B) and with (Gra. C) hydrogen passivation).

Figure 5 shows illuminated *J-V* curves of the best cells from Gra. B and C. The best cell in Gra. C (cell No. 5) i.e. Gra. C-5, has 3.17 $\Omega m^2$ parallel resistance, 502 mV $V_{oc}$ and 67.2% fill factor. In contrast, the best cell in Gra. B i.e. Gra. B-8, without the hydrogen bulk passivation has less $V_{oc}$ and $J_{sc}$ mainly due to the defects in the bulk. To date, we have achieved a cell efficiency of 10.2%, which is the highest efficiency of a 40 $\mu$m thick crystalline silicon solar cell on graphite substrate with a 4 $cm^2$ aperture area.

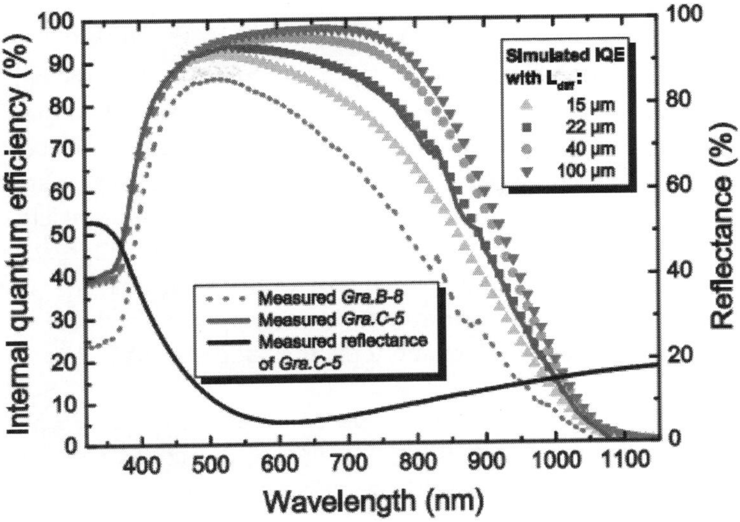


**Figure 6**. The reflectance, measured and simulated IQE curves of the cell Gra. C-5. The measured IQE of the cell Gra. B-8 is plotted for comparison. The bulk lifetime was varied for fitting. The IQE is simulated by PC1D for diffusion lengths with 15, 22, 40, and 100 $\mu$m

**Table 3** . PC1D parameters and simulation results. The parameters device area, base contact, thickness, *p*-type background doping and first rear diffusion were chosen according to cell Gra. C-5. *The parameters were varied for fitting IQE.

| | Parameters | Results | Units |
|---|---|---|---|
| Input | Device area | 4 | $cm^2$ |
| | Base contact | 0.7125 | $\Omega$ |
| | Thickness | 40 | $\mu m$ |
| | *p*-type background doping | $2 \times 10^{16}$ | $cm^{-3}$ |
| | Sheet Resistance | 112 | $\Omega$/square |
| | 1st rear diffusion | $4 \times 10^{18}$ | $cm^{-3}$ |
| | Bulk recombination* | 0.2 | $\mu s$ |
| | Front surface recombination* | $8 \times 10^5$ | cm/s |
| Output | $I_{sc}$ | 0.103 | A |
| | $V_{oc}$ | 567.9 | mV |
| | Efficiency | 10.15 | % |

## 2.2 Quantum efficiency

Figure 6 shows the measured IQE of Gra. B-8 and Gra. C-5, the simulated IQE using PC1D and the reflectance of the Gra. C-5. Gra. B-8 had the shorter diffusion length and lower quantum efficiency than Gra. C-5 due to the surface and bulk recombination without the hydrogen bulk passivation. The simulation parameters are shown in Table 3. The simulation results are in very good agreement with the measured curve. The main difference is that the simulated $V_{oc}$ is about 50 mV higher than the measured one. This may be a result of spatial non-uniformities due to various crystal grains, which cannot be accounted for in 1-dimensional simulations using PC1D.

The anti-reflection coating layers made up of silicon nitride 75 nm thick. The refractive index of the layer is approximately 2. Figure 6 shows that the minimum reflection of the reflectance curve is close to 600 nm, since we fabricated it close to the maximum of the number of incident photons. The $SiN_x$ layers were formed at a low temperature without optimized passivation properties, which cannot reduce the surface recombination in the emitter. Therefore, as shown in the IQE curves, that was assumed to be the main reason why the cell has a spectral response below 90% in the range from 400 nm to 550 nm. In the range from 550 nm to 1000 nm, the IQE goes down gradually due to the bulk and rear recombination loss. In addition, the QE results also reveal that, there are some possibilities for improving the cell performance further, such as by lowering the surface recombination and by optimizing the bulk passivation process.

## 2.3 Diffusion length

The measured average diffusion length of Gra. C-5 is about 120 $\mu$m using the LBiC, however, the value is larger than the cell thickness, since these three wavelengths (976 nm, 951 nm and 846 nm) have the longer penetration lengths than the 40 $\mu$m thick active layers of the cells.

Therefore, we obtained the effective diffusion length of the CSiTF solar cell by using PC1D. The real thickness of the cell was included into the PC1D model. We used this model to fit the measured IQE. The simulated IQE characteristic curves for diffusion lengths were 15, 22, 40, and 100 $\mu$m, respectively. The effective diffusion length was varied for the fitting and is approximately 22 $\mu$m. This result again demonstrates that a surface passivation process and a higher layer quality are necessary for further increasing cell efficiency.

## 3 CONCLUSIONS

In this work, we have presented a new strategy to improve CSiTF solar cell on graphite substrates, using the laser single side contact formation and the hydrogen bulk passivation process at a low temperature. Both methods result in an increase of $V_{oc}$and *FF*. The recombination losses in the bulk and the surface are the main limitations on the cell performance. We achieved 10.2% cell efficiency on an aperture area of 4 $cm^2$.

By comparing different sequences with *J-V* measurement, we also found the best sequence for inclusion of the hydrogen bulk passivation. This significantly improved the cell performance.

We found that one-side contact with LSSC formation is better than front and back contact formation for CSiTF solar cells on graphite substrates. Furthermore, the cell fabrication process can be simplified by making the metallization of all contacts at the same time. The successful application of the LSSC indicates a potential improvement: it is possible to obtain good cell performance at a cost-effective price, while all contacts are on the front side and the impact by substrates is reduced using the LSSC formation. Not only for the CSiTF solar cells on graphite substrates, but we also expect that the LSSC strategy and the best hydrogen bulk passivation process sequence can be applied for the cells based on other high temperature foreign substrates.

So far, a $SiO_x$ intermediate layer is supposed to provide a better quality of surface passivation than a SiC layer [33]. In addition, the batches have not included optimum design of metal grids, light trapping, surface passivation or local diffusion. Thus, the results of both LBiC and QE show possibilities for further improvement.

## ACKNOWLEDGMENTS

The authors gratefully acknowledge funding by the German Federal Ministry for the Environment, Nature Conservation and Nuclear safety

(BMU) under contract No. 0325031B. We thank Dr. Stefan Janz, Fraunhofer ISE Freiburg, for the recrystallization of the silicon layers.

## REFERENCES

1. D. Amkreutz, J. Müller, M. Schmidt, T. Hänel, T.F. Schulze, Prog. Photovolt.: Res. Appl. **19**, 937 (2011)

2. J. Haschke, D. Amkreutz, L. Korte, F. Ruske, B. Rech, Sol. Energy Mater. Sol. Cells **128**, 190 (2014)

3. A.M. Barnett, R.B. Hall, J.A. Rand, C.L. Kendall, D.H. Ford, Sol. Energy Mater. **23**, 164 (1991)

4. T. Kieliba. University of Konstanz, 2006

5. M. Weizman, H. Rhein, J. Dore, S. Gall, C. Klimm, G. Andrä, C. Schultz, F. Fink, B. Rau, R. Schlatmann, Sol. Energy Mater. Sol. Cells **120**, 521 (2014)

6. J. Dore, D. Ong, S. Varlamov, R. Egan, M.A. Green, IEEE J. Photovolt. **4**, 33 (2014)

7. J. Dore, R. Evans, U. Schubert, B.D. Eggleston, D. Ong, K. Kim, J. Huang, O. Kunz, M. Keevers, R. Egan, S. Varlamov, M.A. Green, Prog. Photovolt.: Res. Appl. **21**, 1377 (2013)

8. G. Andrä, F. Falk, Phys. Stat. Sol. C **5**, 3221 (2008)

9. S. Kühnapfel, N.H. Nickel, S. Gall, M. Klaus, C. Genzel, B. Rech, D. Amkreutz, Thin Solid Films **576**, 68 (2015)

10. K.R. Catchpole, M.J. McCann, K.J. Weber, A.W. Blakers, Sol. Energy Mater. Sol. Cells **68**, 173 (2001)

11. R. Brendel, *Thin-film Crystalline Silicon Solar Cells: Physics and Technology* (Wiley-Vch GmbH & Co. KGaA, Weinheim, 2003)

12. J. Poortmans, V. Arkhipov, *Thin Film Solar Cells: Fabrication, Characterization and Applications* (John Wiley & Sons, 2006)

13. C. Becker, D. Amkreutz, T. Sontheimer, V. Preidel, D. Lockau, J. Haschke, L. Jogschies, C. Klimm, J.J. Merkel, P. Plocica, S. Steffens, B. Rech, Sol. Energy Mater. Sol. Cells **119**, 112 (2013)

14. A. Focsa, I. Gordon, J.M. Auger, A. Slaoui, G. Beaucarne, J. Poortmans, C. Maurice, Renew. Energy **33**, 267 (2008)

15. T. Kunz, V. Gazuz, N. Gawehns, I. Burkert, M.T. Hessmann, R. Auer, in *Proceedings of 24th European Photovoltaic Solar Energy Conference and Exhibition, Hamburg, Germany, 2009*, p. 2553

16. S. Janz, P. Löper, M. Schnabel, Mater. Sci. Eng. B **178**, 542 (2013)

17. T. Kunz, M.T. Hessmann, R. Auer, A. Bochmann, S. Christiansen, C.J. Brabec, J. Cryst. Growth **357**, 20 (2012)

18. K. Schillinger, S. Janz, S. Reber, in *IEEE 39th Photovoltaic Specialists Conference, Tampa, USA, 2013*, p. 1784

19. T. Kunz, R. Auer, ZAE Bayern, Report No. FKZ 0325031B, 2013

20. T. Kunz, V. Gazuz, M.T. Hessmann, N. Gawehns, I. Burkert, C.J. Brabec, Sol. Energy Mater. Sol. Cells **95**, 2454 (2011)

21. M.J. Stocks, A. Cuevas, A.W. Blakers, Prog. Photovolt.: Res. Appl. **4**, 35 (1996)

22. C. Hebling, S.W. Glunz, C. Schetter, J. Knobloch, A. Räuber, Sol. Energy Mater. Sol. Cells **48**, 335 (1997)

23. R. Lüdemann, S. Schaefer, C. Schule, C. Hebling, in *Proceedings of the 26th IEEE Photovoltaics Specialist Conference, Anaheim, USA, 1997*, p. 159

24. T. Rachow, M. Ledinsky, S. Janz, S. Reber, A. Fejfar, in *Proceedings of 27th European Photovoltaic Conference and Exhibition, 2012*, p. 2386

25. S. Reber, W. Zimmermann, T. Kieliba, Sol. Energy Mater. Sol. Cells **65**, 409 (2001)

26. T. Kunz, I. Burkert, N. Gawehns, R. Auer, in *Proceedings of the 23rd European Photovoltaic Solar Energy Conference, Valencia, Spain, 2008*, p. 2202

27. B. Gorka, Ph.D. Thesis, Technical University of Berlin, 2010

28. D.A. Clugston, P.A. Basore, in *Proceedings of the 26th IEEE Photovoltaics Specialist Conference, Anaheim, USA, 1997*, p. 207

29. T. Kunz, I. Burkert, R. Auer, A.A. Lovtsus, R.A. Talalaev, Y.N. Makarov, J. Cryst. Growth **310**, 1112 (2008)

30. V. Gazuz, M. Muehlbauer, M. Scheffler, R. Weissmann, R. Auer, in *Proceedings of 21st European Photovoltaic Solar Energy Conference and Exhibition, Dresden, Germany, 2006*, p. 851

31. A. Slaoui, A. Barhdadi, J.C. Muller, P. Siffert, Appl. Phys. A **39**, 159 (1986)

32. F. Granek, A. Weeber, K. Tool, R. Kinderman, P. de Jong, in *Conference Record of the 2006 IEEE 4th World Conference on Photovoltaic Energy Conversion, 2006*, p. 1319

33. J. Dore, S. Varlamov, M.A. Green, IEEE J. Photovolt. **5**, 9 (2015)

# Index